Contents

1 Introduction to Condensed Matter

Condensed matter physics explores the properties of materials containing a large number of atoms. Unlike the study of isolated atoms or molecules, it focuses on emergent phenomena that arise from the collective behavior of these constituent particles—behaviors that are often far more complex and fascinating than the sum of their parts. Think of it like this: a single grain of sand is unremarkable, but a vast expanse of sand, molded by wind and water into dunes and beaches, reveals intricate beauty and structure not inherent in the individual grain. This collective behavior is precisely what defines the realm of condensed matter physics. Understanding these emergent properties requires exploring the intricate interplay between the classical and quantum worlds, a journey that lies at the heart of this book.

The macroscopic properties of condensed matter, such as electrical conductivity, magnetism, and thermal behavior, don't readily stem from simply adding up the individual atomic properties. Instead, these properties arise from collective interactions and subtle quantum effects. The classical approach, while offering valuable insight into some aspects, ultimately fails to accurately describe numerous fundamental phenomena in condensed systems. For instance, the explanation of superconductivity demands the tools of quantum mechanics, as does the understanding of the quantum Hall effect. These examples highlight the crucial need for a blended approach incorporating both classical and quantum perspectives, allowing for a more comprehensive and nuanced understanding. This interdisciplinary approach is essential for tackling complex challenges and making breakthroughs.

The sheer variety of materials encompassed by the field is staggering. From the perfectly ordered crystals of silicon in computer chips to the amorphous structure of window glass, from the superconducting ceramics holding the promise of energy-efficient power grids to the complex alloys underpinning modern aerospace engineering—the range of systems studied is extensive and constantly expanding. And the properties of these materials are equally diverse, leading to exciting applications across various fields. The interplay between structure and behavior plays a crucial role; minute alterations in the arrangement of atoms can dramatically influence the material's macroscopic properties. This underlying relationship is central to much of the research in the field.

Moreover, the development of sophisticated experimental techniques and computational tools has significantly advanced our understanding. Experimental advancements, such as scanning tunneling microscopy, allow scientists to visualize atomic arrangements with exquisite precision, while advanced computational methods like density functional theory provide powerful avenues to predict and understand the complex behavior of materials. These computational tools allow us to model materials with high accuracy, which is crucial when experimental procedures are expensive or impossible to perform. This synergy between experimental observations and theoretical modeling has been instrumental in driving innovation and shaping our knowledge. We now have tools at our disposal to simulate and understand many previously intractable phenomena, opening up countless new avenues for discovery.

Furthermore, this field is not just about understanding existing materials. It is also about designing and creating entirely new materials with properties tailored to specific applications. The ability to manipulate materials at the atomic level, using techniques like epitaxial growth and nanofabrication, opens up exciting possibilities for engineering novel functionalities. For example, the ongoing pursuit of room-temperature superconductors or the development of high-efficiency thermoelectric materials are examples of the transformative potential of this type of research. The practical implications of manipulating matter at the atomic scale have profound impact on several technologies which are in their development phase.

In conclusion, condensed matter physics stands as a vibrant and dynamic field at the forefront of materials science and technology. The study of condensed matter is not just about understanding fundamental physical principles; it is fundamentally linked to technological innovation and has transformative potential across many sectors. Its unique combination of fundamental physics, material science, and engineering opens doors for unprecedented technological advances with impact across several sectors. By understanding the intricacies of condensed matter, we unlock the key to revolutionizing existing technologies and creating entirely new ones. It promises a journey that will enrich and advance our world.

1.1 What is Condensed Matter?

Condensed matter physics focuses on the macroscopic properties of matter arising from the interactions of its constituent atoms and molecules. Unlike isolated atoms or molecules which are studied in atomic or molecular physics, condensed matter systems contain a vast number of particles interacting with each other, resulting in emergent collective behaviors. This complexity is what makes condensed matter physics so rich and challenging, while simultaneously offering immense possibilities for technological innovation. The sheer number of particles makes a direct, particle-by-particle approach computationally infeasible. Instead, we rely on statistical methods and models to analyze and predict the behavior of these systems. Think of a sandcastle—the individual grains of sand are relatively simple, but their collective arrangement creates a complex and intricate structure. Similarly, the macroscopic properties of materials emerge from the interactions of countless atoms.

Imagine a crystalline solid, a highly ordered arrangement of atoms. These atoms vibrate around their equilibrium positions, and these vibrations, collectively called phonons, are critical to understanding the material's thermal properties. The precise arrangement of the atoms within the crystal lattice—cubic, hexagonal, or other geometries—dictates many of the material's properties. These structures are not perfect; defects such as vacancies, interstitial atoms, or dislocations exist, significantly impacting the overall behavior. These irregularities, seemingly minor at the atomic level, can drastically alter the material's strength, conductivity, and other macroscopic characteristics. Therefore, understanding these imperfections is just as crucial as comprehending perfect crystalline order.

The behavior of electrons in a solid is central to understanding its electrical and thermal properties. The simplest model, the free electron model, treats electrons as non-interacting particles moving freely throughout the material. This model provides a basic understanding of electrical conductivity but fails to capture many crucial features, notably the distinction between conductors, semiconductors, and insulators. Band theory offers a more refined picture, taking into account the interaction of electrons with the periodic potential of the crystal lattice. This approach introduces the concept of energy bands—allowed ranges of electron energies separated by band gaps—crucially influencing electrical conductivity and other electronic properties. This helps us understand the differences in how electrons behave in these materials, directly affecting their applications in technology.

Moreover, many materials exhibit magnetic properties. Diamagnetism and paramagnetism represent weak responses to external magnetic fields. The first is a weak repulsion of an external magnetic field, while the latter shows a weak attraction. However, much stronger phenomena like ferromagnetism and antiferromagnetism exist, where strong interactions between electron spins lead to spontaneous magnetization or antiparallel spin ordering, respectively. These interactions, called exchange interactions, are quantum mechanical in nature and are responsible for the technologically important magnetic properties of many materials. The ability to manipulate these magnetic properties is fundamental to the development of countless technologies including magnetic storage devices.

The study of superconductivity, the complete disappearance of electrical resistance below a critical temperature, reveals the profound interplay between classical and quantum phenomena in condensed matter. The Meissner effect, the expulsion of magnetic fields from a superconductor, is a hallmark of this state. The BCS theory provides a microscopic understanding of superconductivity, based on electron-phonon interactions, leading to the formation of Cooper pairs – electron pairs behaving as bosons and exhibiting zero resistance. However, high-Tc superconductors, materials exhibiting superconductivity at much higher temperatures, defy a complete explanation within the BCS framework, revealing much still to be understood. This persistent mystery continues to drive significant research. These superconductors

hold immense potential for applications ranging from energy transmission to medical imaging.
Finally, the quantum Hall effect, observed in two-dimensional electron systems subjected to strong magnetic fields, provides stunning examples of quantum phenomena in condensed matter. The integer and fractional quantum Hall effects reveal the emergence of highly quantized conductance and the existence of fractional charges, reflecting a profound departure from classical physics. These phenomena are closely related to topological insulators, materials with unique edge states allowing for dissipationless current flow along their edges, further highlighting the importance of topology in condensed matter physics. The study of the quantum Hall effect has far-reaching implications for our fundamental understanding of quantum mechanics and its applications in quantum computing and novel devices.

1.2 Classical vs. Quantum Descriptions

Classical physics, with its reliance on continuous variables and deterministic equations, provides a robust framework for understanding many macroscopic phenomena. We can accurately predict the trajectory of a cannonball, the swing of a pendulum, or the flow of a river using classical mechanics and thermodynamics. However, when we delve into the microscopic world of atoms and electrons within condensed matter systems, this classical approach falters. The limitations become starkly apparent when dealing with phenomena like superconductivity, where resistance vanishes entirely, a feat utterly inexplicable using solely classical theories. The scale of the interactions, the sheer number of particles involved, and the nature of their interactions defy classical description. We need a more nuanced, quantum mechanical framework.

Quantum mechanics introduces a radically different perspective. Instead of continuous variables, it deals with quantized energies, momenta, and spins. Electrons, for instance, don't exist in definite locations but rather as probability distributions described by wave functions. Their behavior is governed by probabilistic laws, starkly contrasting with the determinism of classical physics. This probabilistic nature is not a limitation; it's fundamental to the quantum description, revealing insights into the underlying behavior of condensed matter systems. The very existence of energy bands, responsible for distinguishing metals, semiconductors, and insulators, is a direct consequence of quantum mechanics and wave-particle duality, utterly absent in a classical picture.

The transition from a classical to a quantum description isn't always a simple on/off switch. Many properties of condensed matter systems exhibit a fascinating interplay between classical and quantum effects. Consider the specific heat of solids. At high temperatures, the classical equipartition theorem accurately predicts the specific heat capacity, reflecting the average kinetic energy of vibrating atoms. However, at low temperatures, this classical prediction breaks down dramatically. The specific heat drops sharply, approaching zero as the temperature tends to absolute zero. This behavior is explained by the quantization of lattice vibrations—phonons—a quantum mechanical concept directly affecting thermal properties in a measurable, macroscopic way. This illustrates how quantum effects, although inherently microscopic, can profoundly impact macroscopic behavior.

Let's examine another crucial example: the phenomenon of magnetism. While classical electromagnetism explains some magnetic effects, it fails to account for the intrinsic magnetic moment of electrons, their spin. Spin, a purely quantum mechanical property, is responsible for many forms of magnetism, including ferromagnetism, where spontaneous magnetization arises from the parallel alignment of electron spins. This alignment is a direct consequence of quantum mechanical exchange interactions, which are entirely absent in classical models. Consequently, understanding magnetism in materials requires a quantum approach. The intricate relationships between spin, orbital angular momentum, and the surrounding lattice structure require quantum mechanical calculations to even begin to model.

The quantum mechanical description of condensed matter often involves complex mathematical tools and sophisticated computational techniques. Approximations and simplified models are frequently used to handle the immense complexity of many-body systems, and many times they are vital to even begin to explore the system behavior. Density functional theory (DFT), for instance, is a powerful computational technique that helps predict the electronic structure of materials by mapping the complex many-electron problem onto a simpler one involving a single electron moving in an effective potential. This simplification allows for the calculation of various material properties, from band structures to magnetic moments, bridging the gap between theoretical predictions and experimental observations. While approximations are needed, DFT and other similar methods provide invaluable insights into the quantum behavior of condensed matter systems, making them a crucial tool for researchers in the field. The increasing computational power and improved algorithms allow us to address increasingly complex scenarios, improving the accuracy of predictions and helping to refine our understanding of the interplay between quantum and classical descriptions in condensed matter systems.

In conclusion, understanding condensed matter physics necessitates a sophisticated understanding of both classical and quantum mechanics. While classical physics provides a valuable framework for describing macroscopic behavior, quantum mechanics is essential for capturing the fundamental microscopic interactions that drive macroscopic properties. Many phenomena observed in condensed matter systems are a direct consequence of the quantum nature of matter and defy any classical explanation. The interplay between classical and quantum descriptions is a central theme in this area, and mastering both frameworks is crucial for any researcher, physicist, and chemist pursuing in-depth exploration of condensed matter physics. The combination of these two perspectives, combined with computational techniques like DFT, provides a powerful toolkit for understanding and predicting the behavior of these fascinating materials.

2 Crystalline Solids

Crystalline solids form the bedrock of much of condensed matter physics. Their regular, repeating structures, unlike the amorphous jumble of liquids or gases, allow for a precise and powerful theoretical treatment. This predictability stems from the underlying lattice, a three-dimensional arrangement of atoms, ions, or molecules. Understanding these lattices is paramount; it unlocks the secrets of numerous material properties. We start by analyzing the simplest cubic lattices: simple cubic, body-centered cubic, and face-centered cubic, each distinguished by its unique arrangement of lattice points and the resulting symmetry. These arrangements aren't mere abstractions; they directly influence properties like density and packing efficiency. Moving beyond the cubic structures, we can explore more complex lattices like hexagonal close-packed and others, each displaying different symmetries and impacting the material's response to external forces and fields. The consequences of these structural differences are profound, leading to a diverse range of mechanical, thermal, and electrical behaviours. The impact of lattice structure on macroscopic material properties forms a fascinating interplay between the microscopic world of atoms and the macroscopic world of materials.

The atoms in a crystal lattice aren't stationary; they vibrate around their equilibrium positions. These vibrations, far from being random noise, are quantized, meaning they exist in discrete energy levels. These quantized vibrations are called phonons, and they are crucial for understanding many of the thermal properties of solids. The phonon concept, rooted in quantum mechanics, allows for the calculation of heat capacity, thermal conductivity, and thermal expansion—properties that determine how a material responds to changes in temperature. Understanding phonon dispersion relationships—the relationship between phonon frequency and wave vector—is particularly essential for this task, as it reveals how phonons interact and transport energy through the lattice. Moreover, the study of phonons provides a powerful method to investigate the interactions between lattice vibrations and other phenomena within the material, for example, their influence on electron transport and superconductivity. Therefore, mastering the phonon concept is crucial for any serious study of crystalline solids.

Imperfections, while seemingly undesirable, are often intrinsic to crystalline solids and play a crucial role in determining their properties. These defects disrupt the perfect periodicity of the lattice, altering the material's behavior in significant ways. Point defects, like vacancies (missing atoms) and interstitials (extra atoms), alter local lattice structure, affecting electrical conductivity and diffusion. Line defects, or dislocations, are essentially imperfections in the lattice structure that arise from misalignments of crystal planes. Dislocations affect the mechanical strength of a material, making it either stronger or weaker depending on the type and density of the dislocations present. Planar defects, such as grain boundaries and stacking faults, represent regions where the crystallographic order is disrupted across two-dimensional planes. These disruptions impact properties like diffusion and fracture behavior. Understanding these defects is essential, not merely for understanding the imperfections, but for manipulating them to engineer materials with desired properties. Controlling defects allows for the

tuning of material properties for specific applications.
Beyond simple defects, crystalline solids often display more complex imperfections. These include extended defects like stacking faults, where layers of atoms are incorrectly stacked, leading to altered crystal structure and properties. The influence of stacking faults on mechanical and electronic properties is substantial. Grain boundaries, regions separating crystallites (small crystals) with different orientations, are ubiquitous in polycrystalline materials. They act as barriers to dislocation motion, influencing the overall material strength. Further complexities arise from impurities and dopants intentionally introduced to fine-tune a material's characteristics. For example, the addition of specific impurities, or doping, can drastically change electrical conductivity—a fundamental principle used in semiconductor technology. Therefore, considering these complex defects is key for a comprehensive understanding of crystalline solid behavior, bridging the gap between ideal theory and the complexities of real-world materials. This comprehension is fundamental to materials science and engineering, enabling the design and synthesis of novel materials with specific functionalities.

2.1 Lattice Structures

Crystalline structures underpin the macroscopic properties of many solids. Their seemingly simple arrangement of atoms, ions, or molecules dictates everything from electrical conductivity to the material's response to light. Understanding lattice structures is fundamental to comprehending the behavior of condensed matter systems, bridging the gap between microscopic interactions and macroscopic observations. We begin by examining the basic building blocks: unit cells. These smallest repeating units define the overall symmetry and periodicity of the entire crystal lattice.

The simplest lattice structures are the cubic systems, encompassing simple cubic (sc), body-centered cubic (bcc), and face-centered cubic (fcc) arrangements. These structures, while seemingly straightforward, demonstrate the profound impact of atomic packing on material properties. For example, the denser packing in fcc structures often leads to higher densities and different mechanical strengths compared to sc structures. The relative arrangement of atoms directly influences the interactions between them, a crucial factor in determining various physical properties.

Moving beyond cubic systems, we encounter a wider array of lattice structures classified by their symmetry properties. These structures exhibit different levels of rotational and translational symmetry, influencing how they interact with external fields and the types of defects they can accommodate. The hexagonal close-packed (hcp) structure, for instance, displays a distinct arrangement of atoms compared to cubic lattices, leading to unique mechanical and optical properties. The detailed classification and analysis of these crystal systems require the use of crystallographic notation and group theory, tools that elegantly describe the spatial symmetries inherent in these structures.

Within each lattice structure, the atoms may be arranged in specific patterns, influencing the macroscopic properties of the material. The locations of atoms within the unit cell determine the interatomic distances and angles, which in turn influence the strength of bonding. Stronger bonding generally leads to materials that are harder, have higher melting points, and exhibit different elastic properties. Precise measurements using techniques like X-ray diffraction are essential for determining the exact atomic positions and characterizing the overall structure.

Understanding these fundamental crystallographic principles is paramount for predicting the behavior of materials. The knowledge of lattice structures provides a foundation for understanding other crucial concepts in condensed matter physics, such as phonons, which are quantized lattice vibrations. Furthermore, understanding the lattice structure is vital for analyzing the effects of crystal defects, which can significantly alter the properties of a material. It serves as a bridge connecting the microscopic world of atoms and molecules to the macroscopic world of observable physical properties. The precise arrangement of atoms creates the stage for the quantum phenomena explored later in the book.

The interplay between the classical description of lattice structures and the quantum behavior of electrons within these structures is a recurring theme in condensed matter physics. While the lattice itself can often be adequately described classically, the behavior of electrons is fundamentally governed by quantum mechanics. This interplay is responsible for a wide range of fascinating phenomena, from superconductivity to the quantum Hall effect, which are discussed extensively in subsequent chapters. Comprehending the framework of crystal structures lays a solid base for appreciating these more complex quantum phenomena.

The determination of crystal structures typically involves sophisticated techniques. X-ray diffraction, neutron diffraction, and electron diffraction are powerful tools used to investigate the arrangement of atoms within a crystalline material. These methods exploit the wave nature of radiation or electrons to probe the periodic structure of the lattice, generating diffraction patterns that directly reflect the lattice parameters and symmetry. Analyzing these patterns provides invaluable insights into the structure and ordering within the crystalline material. These techniques form the cornerstone of experimental crystallography and allow for the precise determination of lattice parameters, atom positions, and crystal symmetry, forming the base for many theories and models presented throughout the book.

2.2 Phonons and Vibrations

Imagine a crystal lattice, a seemingly rigid structure. Yet, at a microscopic level, a constant ballet of atomic vibrations unfolds. These aren't random jiggles; they're collective oscillations, waves propagating through the lattice, carrying energy and momentum. We call these quantized vibrations phonons, a cornerstone of understanding many properties of solids. Think of them as the "sound" of the crystal lattice, but on a scale far smaller than anything audible to the human ear. Their energy is quantized, meaning it exists in discrete packets, akin to the photons of light or the electrons in an atom. This quantization arises directly from the quantum mechanical nature of the system, an essential feature that distinguishes it from a purely classical description of atomic vibrations. The energy of a phonon is directly proportional to its frequency, a fundamental relationship mirroring that found in other quantum systems, showcasing the deep interconnectedness of physical principles across diverse phenomena.

The interaction of phonons with other quasiparticles, such as electrons, is crucial in determining the material's macroscopic behavior. For instance, the electrical conductivity of a metal is heavily influenced by electron-phonon scattering. At low temperatures, these scattering events are minimal, allowing electrons to move more freely, thus exhibiting high conductivity. However, as temperature increases, the vibrational amplitude of phonons rises, resulting in more frequent scattering events, ultimately leading to a decrease in conductivity. This elegant connection between microscopic lattice vibrations and macroscopic electrical properties demonstrates the power of the phonon concept in explaining observed phenomena. A thorough understanding of this interaction is fundamental for developing materials with desired electrical properties, a key aspect in advancing technology. Furthermore, phonons play a key role in determining a material's thermal conductivity. Their ability to carry heat significantly influences a material's response to temperature changes. Materials with high phonon thermal conductivity, for example, are excellent heat sinks, finding extensive applications in electronics to dissipate excess heat generated by components. Conversely, materials with low phonon thermal conductivity are desired for thermal insulation applications.

Consider the specific case of superconductivity. In the BCS theory, which provides a microscopic explanation for conventional superconductivity, electron-phonon interactions are the driving force behind the formation of Cooper pairs – the electron pairs responsible for the dissipationless flow of electric current. The exchange of phonons between electrons creates an attractive interaction that overcomes the Coulomb repulsion, leading to the formation of these bound pairs. This intricate interplay is a striking example of how seemingly simple lattice vibrations can have profound consequences on the macroscopic behavior of a material, giving rise to a phenomenon as extraordinary as superconductivity. This highlights the crucial role of phonon dynamics in shaping the rich variety of physical phenomena observed in condensed matter systems.

Moreover, phonons are not solely confined to perfect crystalline structures. The presence of defects and imperfections in the crystal lattice alters the phonon spectrum, affecting its thermal and electrical properties. For instance, impurities can scatter phonons, reducing the thermal conductivity. This effect is exploited in the design of thermoelectric materials – materials that can efficiently convert heat into electrical energy or vice-versa. The ability to control phonon transport through careful design of materials with specific defects is a vibrant area of research, offering potential for energy-efficient technologies and advanced thermal management systems. Beyond simple impurities, more complex

defects, such as dislocations and grain boundaries, drastically alter the phonon behavior, leading to significant changes in the macroscopic material properties. The study of how phonons are influenced by these defects requires advanced computational techniques and provides insights crucial for optimizing material performance.
Studying phonons involves various experimental techniques. Inelastic neutron scattering is a powerful tool used to directly probe the phonon dispersion relations – the relationship between phonon energy and momentum. This technique utilizes the interaction of neutrons with the lattice vibrations, allowing for a detailed mapping of the phonon spectrum. Raman spectroscopy, another important technique, provides information about the vibrational modes of the crystal lattice by analyzing the scattering of light. This non-destructive technique is particularly useful for studying the vibrational properties of surfaces and interfaces, where the phonon behavior can differ significantly from that of the bulk material. Together, these experimental techniques, coupled with theoretical calculations, allow for a comprehensive understanding of phonon dynamics in condensed matter systems, providing critical insights for designing materials with specific and tailored properties. The knowledge and understanding of phonons and their interaction with other quasiparticles within the condensed matter system is of fundamental importance for the development of advanced materials and the future of various technological applications. The ability to manipulate and control phonon behavior offers a wide range of possibilities for technological advancements.

2.3 Crystal Defects

Crystal defects, imperfections in the otherwise regular arrangement of atoms in a crystal lattice, profoundly impact a material's properties.

They're not mere anomalies; understanding them is crucial for tailoring materials to specific applications. Think of it like this: a perfectly formed diamond, flawless and pristine, represents the theoretical ideal, strong and transparent. But real-world diamonds, even the most valuable ones, contain defects that influence their color, clarity, and even strength.

These imperfections come in various forms. Point defects involve a single atom or a few atoms. Vacancies, where an atom is missing from its lattice site, are among the simplest. Interstitials, conversely, involve extra atoms squeezed into the spaces between lattice sites. Substitutional defects occur when an atom of a different element replaces a host atom. These variations in atomic arrangement alter the local electronic structure and can lead to significant changes in the material's overall behavior. For instance, doping silicon with phosphorus introduces extra electrons, transforming the insulator into an n-type semiconductor crucial for electronics.

Line defects, also known as dislocations, are one-dimensional imperfections. Imagine a slipped plane of atoms within the crystal lattice; this disruption in the atomic arrangement affects material properties like plasticity and strength. Edge dislocations resemble an extra half-plane of atoms inserted into the crystal, creating strain around the defect. Screw dislocations, on the other hand, create a spiral ramp in the lattice structure, dramatically altering the crystal's mechanical properties. Understanding dislocation movement and interaction is vital in controlling material strength and ductility, influencing the design of everything from bridges to microchips.

Planar defects are two-dimensional imperfections extending across the crystal. Grain boundaries separate regions of differently oriented crystallites in polycrystalline materials. These boundaries act as barriers to dislocation motion, influencing the material's strength and other properties. Stacking faults disrupt the regular sequence of atomic planes within the crystal structure, leading to alterations in the material's electrical and magnetic properties. Twin boundaries are a specific type of planar defect where a mirror symmetry exists across the boundary plane. Their presence can significantly impact the overall mechanical behavior of the material.

Finally, we have volume defects, three-dimensional imperfections that extend throughout a significant region of the material. These include pores, voids, and inclusions of foreign phases. These defects can weaken the material and affect its properties considerably. Pores, for example, reduce the material's density and overall strength, increasing its susceptibility to fracture. Inclusions, on the other hand, could be beneficial or detrimental. If the included phase enhances the material's properties, it can be beneficial. If not, it can lead to material failure. The influence of these defects necessitates careful control of material processing techniques to ensure the desired material performance.

The study of crystal defects is not just an academic exercise; it's a critical aspect of materials science and engineering. By controlling the type and concentration of defects, we can tailor the properties of materials for specific applications. For instance, the intentional introduction of defects is used to enhance the hardness of certain metals, to alter the electrical conductivity of semiconductors, or to manipulate the optical properties of crystals. Careful consideration of defects is fundamental to the design and fabrication of high-performance materials used in a wide array of technologies, from advanced electronics to aerospace components. Indeed, the pervasive influence of crystal defects underlines the complexities and intricate beauty of condensed matter systems and the endless possibilities inherent in understanding and controlling them.

3 Electronic Structure

The free electron model, a cornerstone of understanding electronic structure, offers a simplified yet insightful perspective. Imagine a crystalline solid as a box containing electrons, free to roam without significant interactions. This approximation, while neglecting the intricate potential landscape created by the atomic nuclei, allows us to grasp fundamental concepts like Fermi energy, the highest occupied energy level at absolute zero, and Fermi surface, the boundary defining occupied states in momentum space. Its predictions for heat capacity and electrical conductivity, though imperfect, lay the groundwork for more sophisticated models. The limitations become apparent when considering the discrepancies between theoretical predictions and experimental observations in real materials, highlighting the need for a more refined approach. The simplicity, however, makes it an invaluable introductory concept, offering a stepping stone to the complexities of band theory.

Band theory refines the free electron model by explicitly incorporating the periodic potential of the crystal lattice. This potential, arising from the positively charged atomic nuclei, profoundly affects the electron wavefunctions, causing them to be modified from the simple plane waves of the free electron model. Instead of a continuous spectrum of energy levels, electrons occupy distinct energy bands, separated by energy gaps. The filling of these bands determines the material's electrical properties. Metals, with partially filled bands, exhibit high conductivity due to the availability of easily excited electrons. Insulators, with fully filled valence bands and a large band gap separating them from the empty conduction band, resist the flow of current. Semiconductors, possessing a small band gap, exhibit intermediate conductivity, often strongly influenced by temperature and doping. Understanding the interplay between band structure and material properties is crucial in designing and tailoring materials for specific applications, from transistors to solar cells. The calculation of these band structures, often relying on complex computational techniques, reveals intricate details that significantly impact material behavior.

Semiconductors, residing in the fascinating realm between conductors and insulators, display a unique sensitivity to external factors, making them the cornerstone of modern electronics. The relatively small energy gap between their valence and conduction bands allows for easy electron excitation, leading to varying conductivity depending on temperature, light exposure, or doping. Doping, the introduction of impurity atoms into the semiconductor lattice, provides a means to precisely control the carrier concentration, enhancing conductivity in n-type semiconductors (with excess electrons) and p-type semiconductors (with electron "holes" or missing electrons). The p-n junction, formed by combining p-type and n-type semiconductors, forms the basis of diodes and transistors, fundamental components of electronic circuits. The control and manipulation of electron flow in these devices underpin the technological revolution driven by semiconductors, a direct testament to our understanding and manipulation of electronic structure. Furthermore, ongoing research continues to refine semiconductor technologies, pushing the boundaries of device miniaturization and performance.

Metals, characterized by their high electrical conductivity, owe this property to their partially filled energy bands. The availability of numerous electrons near the Fermi level allows for facile current flow under an applied electric field. The microscopic origins of this conductivity are directly tied to the electronic structure, specifically the availability of delocalized electrons in the conduction band. These electrons are not bound to specific atoms but move freely within the crystal lattice, a key difference from insulators where electrons are tightly bound within localized atomic orbitals. The ability of these delocalized electrons to carry current is also deeply linked to the scattering of electrons from imperfections in the crystal lattice, as impurities, lattice vibrations (phonons) and other defects restrict the electron's motion, affecting the conductivity. Furthermore, the interaction of electrons with lattice vibrations, giving rise to a temperature dependence of resistivity, contributes to the complex nature of metallic conduction. The ability to precisely engineer and manipulate the crystalline structure of metals, thus controlling their electronic properties, is essential in the design of advanced alloys and materials.

Density Functional Theory (DFT), a powerful computational method, has revolutionized the way we investigate electronic structure in complex materials. DFT leverages the concept that the ground state properties of a many-electron system can be determined by solving equations for a single electron moving in an effective potential. This effective potential includes the external potential from the atomic nuclei and a term accounting for the electron-electron interactions. The simplification is remarkable, reducing the complexity of solving Schrödinger's equation for many interacting electrons to a more manageable problem involving a single electron. While approximations are necessary, DFT often delivers accurate results for diverse properties, ranging from energy band structures to magnetic moments. Its wide applicability, coupled with advances in computational power, has made DFT an indispensable tool for materials science and condensed matter physics, enabling predictions of materials properties before experimental synthesis, and significantly accelerating materials discovery and design. Continued development and refinement of DFT-based methods are pushing the boundaries of what we can compute and understand about complex condensed matter systems.

3.1 Free Electron Model

The free electron model, despite its simplicity, provides a foundational understanding of metallic conductivity. Imagine a metal as a box containing a vast number of electrons, completely free to roam within its confines, unaffected by the positive ions forming the lattice. These electrons behave like a gas, obeying Fermi-Dirac statistics, a quantum mechanical rule dictating electron distribution at different energy levels. This model successfully explains several key properties of metals, including their high electrical and thermal conductivities, offering a starting point for more complex models. The simplicity allows for straightforward calculations, crucial for beginning learners grasping fundamental concepts.

This model's success lies in its ability to capture the essence of metallic behavior. Electrons, considered free from the influence of the crystal lattice, respond readily to external electric fields, resulting in the significant current flow characteristic of metals. The temperature dependence of resistivity, another key metallic property, is also explained by the model, albeit with some limitations. As temperature increases, electron scattering from lattice vibrations increases, hindering their movement and thus raising resistivity—a direct consequence of the electron-phonon interaction.

However, the free electron model also has limitations. It fails to accurately predict many observed phenomena. For instance, it struggles to account for the different magnetic properties of various metals or accurately predict the electronic specific heat of a solid. These discrepancies arise because the model ignores the periodic potential produced by the positively charged ion cores. It fails to address the interaction between electrons and the lattice, neglecting the crucial role played by electron-electron interactions and the band structure of materials, which plays a significant role in determining electrical properties. It is crucial to remember its place as a simplified initial approach, rather than a comprehensive representation of real metallic behaviour.

To reiterate, the free electron model provides a useful, simplified picture of electron behavior in metals. It uses a simplified picture of electrons moving freely in a box, a model which neglects electron-electron interactions and assumes a constant potential energy within the metal. While neglecting the potential from the positive ions is a massive simplification, it nevertheless offers a tractable starting point for understanding certain aspects of metallic behavior.

Despite its simplification, the model yields valuable insights. Calculating the Fermi energy, a crucial parameter representing the highest occupied energy level at absolute zero temperature, is quite straightforward within this framework. This energy level provides an indication of the kinetic energy of electrons in a metal, demonstrating how the electron gas model correlates with experimental observations such as electronic specific heat. The model also helps in understanding the concept of the Fermi surface, a boundary separating occupied and unoccupied states in momentum space, and lays the foundation for exploring more sophisticated approaches.

The free electron gas model offers a conceptual entry point to solid-state physics. Its mathematical simplicity allows for a straightforward understanding of many fundamental concepts such as Fermi-Dirac statistics, Fermi energy, and the density of states. However, its deficiencies emphasize the necessity of using more complete models that incorporate the periodic potential of the ionic lattice, like the nearly free electron model or the full tight-binding approximation. These more complex models will subsequently build upon this fundamental understanding, gradually refining our comprehension of electron behavior in solids.

The free electron model's value lies not in its flawless accuracy but in its pedagogical role. This simplification enables a comprehensive grasp of essential concepts before tackling more sophisticated and nuanced models. It allows one to conceptualize the quantum mechanical behavior of electrons in a metal, setting the stage for the study of band theory and the intricacies of electronic structure. Through this simplified starting point, students can then learn to refine and enhance their understanding to accommodate the more complex aspects of reality.

In conclusion, the free electron model serves as an invaluable introduction to the electronic properties of metals, providing a simple and intuitive way to understand fundamental concepts. Though it simplifies the complex reality of electron behavior in solids by neglecting the periodic potential of the lattice, its predictive power for certain properties such as electrical conductivity makes it an essential tool for introductory studies in condensed matter physics. It provides a strong foundation for the development of more sophisticated models and for a deeper appreciation of the intricate interplay between electrons and the lattice structure in determining the properties of materials. Remembering its limitations and understanding when it's an appropriate tool, this model is a critical starting point for any aspiring condensed matter physicist.

3.2 Band Theory

Electrons in a solid don't behave like free particles; they interact with the periodic potential created by the atomic nuclei. This interaction fundamentally alters their energy levels, causing them to form bands of allowed energies separated by gaps where no electron states exist. Understanding this band structure is crucial for explaining a material's electrical, optical, and thermal properties. The simplest model, the nearly-free electron approximation, treats the periodic potential as a weak perturbation to the free electron gas, leading to the formation of energy bands with slight energy gaps at the Brillouin zone boundaries. The Kronig-Penney model offers a more rigorous, though still simplified, approach. It models the periodic potential as a series of rectangular barriers, providing an analytic solution demonstrating the formation of energy bands and forbidden gaps. This model, while idealized, effectively illustrates the fundamental physics driving the band structure formation: the wave nature of electrons and their interference within a periodic potential. These interference patterns, constructive and destructive, dictate the allowed and forbidden energy ranges, shaping the material's characteristics.

However, calculating the band structure for real materials is far more complex. Tight-binding models, for instance, offer a different perspective. They begin by considering the atomic orbitals of the constituent atoms. Then, through linear combinations of these atomic orbitals, they construct molecular orbitals that extend throughout the crystal lattice, forming the energy bands. These calculations consider the overlap of atomic orbitals and the interactions between electrons, providing a more accurate description, particularly for materials with strongly localized electrons. The complexity increases significantly when dealing with intricate crystal structures and strong electron-electron interac-

tions, often requiring advanced computational techniques.
The resulting band structure isn't merely an abstract concept; it directly dictates a material's properties. For example, in metals, the highest occupied band—the valence band—is partially filled. This allows electrons to easily move under an applied electric field, leading to high electrical conductivity. The situation is drastically different in insulators. Here, the valence band is completely filled, and a large energy gap, the band gap, separates it from the empty conduction band. A significant energy input is needed to excite electrons into the conduction band, resulting in low conductivity.
Semiconductors occupy an intermediate position. They possess a smaller band gap than insulators, allowing for some electron excitation into the conduction band at higher temperatures or under illumination. Doping—introducing impurities into the crystal lattice—further modulates their conductivity. Adding donor impurities provides extra electrons to the conduction band, creating n-type semiconductors; conversely, adding acceptor impurities creates holes in the valence band, forming p-type semiconductors. These are the fundamental principles behind the electronics industry, relying on carefully engineered band structures of semiconductor materials.
Beyond basic conductivity, band structure also influences other properties. Optical absorption, for example, strongly depends on the band gap. Materials absorb photons with energies exceeding the band gap, exciting electrons from the valence to the conduction band; photons with lower energies are transmitted. This dictates the color of the material. Similarly, thermal conductivity is partially determined by the band structure, as electron mobility within the bands plays a role in heat transfer. The specifics of band structure, including band dispersion and effective mass, influence many thermodynamic and transport properties. Therefore, the detailed understanding of a material's band

structure is crucial for predicting and manipulating its macroscopic properties.

Furthermore, understanding the intricacies of band structure opens doors to discovering novel materials with tailored properties. By designing materials with specific band structures, scientists can engineer materials with desired optical, electrical, or magnetic properties. Computational methods, like density functional theory, play a crucial role in predicting the band structure of complex materials, guiding experimental efforts. The continuous exploration and advancement in materials science and condensed matter physics are heavily reliant on the accurate determination and manipulation of band structures. The insights gained from band theory extend far beyond simple categorization of metals, semiconductors, and insulators; it provides a fundamental framework for understanding the behavior of electrons in solids, enabling the design and development of advanced materials with unprecedented functionalities. This understanding is fundamental not only for creating better transistors and integrated circuits but also for pushing the frontiers of technological innovation across various fields.

3.3 Semiconductors and Insulators

Semiconductors and insulators represent fundamentally different responses to the application of an electric field. Unlike metals, where electrons flow freely, creating high conductivity, these materials exhibit a distinct behavior governed by the electronic band structure, a concept central to understanding their unique properties. This difference stems from the energy gap—the forbidden energy region separating the valence band (filled electron states) and the conduction band (empty electron states). In insulators, this band gap is substantial, requiring a large energy input to excite electrons into the conduction band and initiate current flow. Conversely, semiconductors possess a much smaller band gap, allowing for some electrons to transition to the conduction band even at modest temperatures or with the addition of impurities. The magnitude of this band gap directly correlates with the material's conductivity. Insulators, with their large band gaps (typically exceeding 4 electron volts), act as excellent electrical insulators. Their electrons are tightly bound to their atoms, strongly resisting external electric fields. Examples include diamond, a classic insulator with a wide band gap resulting from the strong covalent bonds between carbon atoms, and various oxides like silicon dioxide (SiO2), widely used as an insulator in electronic devices due to its exceptional dielectric strength and stability. These materials form the basis of many electrical components, preventing unwanted current leakage and ensuring proper circuit function.

Semiconductors, on the other hand, exhibit an intermediate conductivity that falls between conductors and insulators. Their smaller band gaps (typically ranging from 0.1 to 2 electron volts) enable a degree of electron mobility under certain conditions. Silicon, the cornerstone of modern electronics, is a prime example. At room temperature, a small fraction of electrons possess sufficient thermal energy to overcome the band gap and jump to the conduction band, leading to a measurable conductivity. This intrinsic conductivity can be significantly altered through doping—intentionally introducing impurities that either donate extra electrons (n-type doping) or create "holes" (electron vacancies) in the valence band (p-type doping).

Doping drastically modifies the semiconductor's conductivity, transforming it from an intrinsically poor conductor into a highly efficient electron or hole carrier. The precise control over this doping process underpins the remarkable versatility of semiconductors in electronic applications. By strategically doping regions of a silicon crystal, one can engineer complex devices like transistors and integrated circuits, the fundamental building blocks of modern computing and electronics. The ability to fine-tune the conductivity through doping makes semiconductors indispensable in countless technologies.

The band structure isn't merely an abstract concept; it's a direct reflection of the atomic arrangement and bonding characteristics within the material. In insulators, strong covalent or ionic bonds create localized electron states, resulting in a wide band gap. The valence electrons are tightly bound to their respective atoms, making it extremely difficult for them to move freely. In semiconductors, weaker bonds allow for a higher degree of electron delocalization, leading to a narrower band gap and increased conductivity under specific conditions like increased temperature or doping. Understanding this interplay between bonding, band structure, and conductivity is key to designing and developing new materials with tailored electrical properties.

Moreover, the temperature dependence of conductivity in semiconductors highlights the quantum mechanical nature of these materials. At absolute zero, semiconductors behave like perfect insulators—all electrons reside in the valence band. As temperature increases, more electrons gain enough thermal energy to overcome the band gap, increasing conductivity exponentially. This temperature dependence contrasts sharply with the relatively constant conductivity of metals, where electron mobility is largely unaffected by temperature variations within a certain range. The sensitivity of semiconductor conductivity to temperature is exploited in numerous applications, like temperature sensors and thermistors. This underlines the significant role of quantum mechanics in governing the behavior of these materials. Careful consideration of these effects is crucial for device design and performance optimization.

Finally, the study of semiconductors and insulators extends beyond their electrical properties. They also exhibit unique optical characteristics due to their band gaps. Photons with energy exceeding the band gap can excite electrons from the valence to the conduction band, a process known as photoconductivity. This underlies the operation of photodetectors and solar cells, which convert light energy into electrical current. Conversely, materials with specific band gaps can absorb certain wavelengths of light while transmitting others, leading to applications in optical filters and colored materials. The detailed understanding of the interplay between electronic band structures and optical properties unlocks further technological advancements in fields like optoelectronics and photonics, illustrating the vast practical implications of research in this area. The continued investigation of novel semiconductor and insulator materials, along with refined control over their fabrication processes, remains crucial for pushing the boundaries of technological innovation.

3.4 Metals and Conductivity

Electrons, the fundamental building blocks of electrical conductivity in metals, behave quite differently than in insulators or semiconductors. Unlike their counterparts in insulators where electrons are tightly bound to their respective atoms, metal electrons exist in a sea of delocalized states, free to roam throughout the material's lattice structure. This freedom of movement is the key to their excellent electrical conductivity. The "sea" is a simplified representation, of course. A more precise description involves quantum mechanics. The electrons occupy a range of energy levels, forming what's known as a conduction band. This band is partially filled, which means there are many unoccupied states readily available for electrons to jump into under the influence of an external electric field.

This ability to easily transition between energy states is crucial. It allows a continuous flow of electrons, thereby enabling the conduction of electric current. Consider an electric field applied across a metallic wire: the electrons respond almost instantaneously. They accelerate, and this directional movement constitutes the electric current. The ease with which electrons move within this conduction band distinguishes metals from other materials. In insulators, the valence band (where electrons are normally found) is completely filled, and there's a substantial energy gap separating it from the conduction band. This means significant energy is needed to excite electrons into the conduction band to initiate electrical flow.

The concept of Fermi energy provides a more nuanced view. The Fermi energy defines the highest occupied energy level of electrons at absolute zero temperature. In metals, the Fermi energy lies within the conduction band, signifying the presence of numerous readily available energy states for electrons to occupy and move. This characteristic contributes to high electrical conductivity at room temperatures and even low temperatures. At absolute zero, the Fermi energy perfectly separates occupied and unoccupied states; as temperature increases, some electrons gain energy and move to higher states, slightly modifying conductivity.

Now, let's dive into the influence of temperature on conductivity. At higher temperatures, the increased thermal vibrations of the lattice atoms interfere with the smooth movement of electrons. These vibrations, essentially phonons, act as scattering centers, impeding electron flow and thus reducing conductivity. This effect is inversely proportional to temperature for many metals, meaning conductivity decreases as temperature rises. This relationship can be well approximated by the Matthiessen's rule, although real-world situations often show slight deviations from it.

However, the impact of impurities should not be underestimated. Even minuscule amounts of impurities within the metallic structure can dramatically alter its electrical conductivity. These impurities introduce lattice distortions, disrupting the electron's free path and scattering them, which directly affects conductivity. The effect is cumulative: more impurities, greater scattering, lower conductivity. The impact of both temperature and impurity levels underscores the complex interplay of factors affecting a metal's electrical behavior.

Superconductivity, a remarkable phenomenon, represents an extreme case where the electrical resistance of certain materials completely vanishes below a critical temperature. In these materials, electrons form pairs called Cooper pairs, mediated by interactions with lattice vibrations (phonons). These pairs move through the lattice without any resistance, making superconductors excellent conductors of electricity with almost no energy loss. The formation of these Cooper pairs fundamentally alters the electronic structure and eliminates the scattering mechanisms that normally limit conductivity. It is a quantum phenomenon, distinct from simple metallic conductivity.

Furthermore, the concept of mean free path plays a vital role in understanding metal conductivity. The mean free path refers to the average distance an electron travels between collisions with lattice atoms or impurities. A longer mean free path directly translates to higher conductivity, as electrons can traverse greater distances without their movement being interrupted. The mean free path significantly reduces with increased temperature and impurity concentration.

Finally, let us not forget the electrical resistivity, the inverse of conductivity. High resistivity indicates low conductivity, and it reflects the extent to which a material impedes the flow of electric current. Resistivity is influenced by various factors; temperature, impurities, and crystal structure are just a few. Precise measurement of resistivity provides valuable insights into the material's electronic structure and the strength of electron-lattice and electron-impurity interactions. Understanding these concepts offers a strong foundation for further exploration into advanced aspects of metallic behavior.

3.5 Density Functional Theory

Density functional theory (DFT) offers a powerful computational approach to predicting the electronic structure of materials. Unlike simpler models like the free electron theory, DFT tackles the complex many-body problem of interacting electrons within a solid, making it invaluable for understanding diverse properties. At its core, DFT cleverly circumvents the direct calculation of the many-electron wavefunction, a task computationally impossible for systems with more than a few electrons. Instead, it focuses on the electron density, a significantly simpler quantity. This ingenious shift in perspective allows for accurate predictions of ground-state properties, including energy levels, charge distributions, and magnetic moments, even for systems containing thousands of atoms. Its application spans a vast array of materials science, from designing novel catalysts to predicting the properties of advanced semiconductor devices.

The theoretical foundation of DFT rests on two key theorems proven by Hohenberg and Kohn. The first establishes that the ground-state electron density uniquely determines the external potential, and consequently, all other ground-state properties of the system. This implies a one-to-one mapping between the electron density and the full many-body wavefunction. The second theorem guarantees that the ground-state electron density minimizes the total energy functional. This energy functional incorporates kinetic energy, electron-electron interactions, and the interaction with the external potential, all expressed as functionals of the electron density. While the exact form of the kinetic energy functional remains unknown, Kohn-Sham theory introduces a clever workaround, replacing the interacting electron system with a fictitious system of non-interacting electrons moving in an effective potential.

This effective potential, a key component of the Kohn-Sham equations, incorporates both the external potential and a term accounting for the electron-electron interaction. This crucial term, the exchange-correlation potential, captures the complexities of the many-body interactions and is often approximated. Various approximations exist, each with its own strengths and limitations, ranging from the simple local density approximation (LDA) to more sophisticated generalized gradient approximations (GGA). The choice of approximation greatly impacts the accuracy of the DFT calculation. LDA works best for systems with slowly varying electron densities, while GGAs offer improved accuracy for systems with more localized electrons and inhomogeneous densities. Hybrid functionals that incorporate a fraction of exact exchange further enhance accuracy, particularly for systems with strong electron correlation.

Implementing DFT calculations involves solving the Kohn-Sham equations, typically using numerical methods. These equations are self-consistent; meaning the solution iteratively refines itself until convergence. The initial guess for the electron density is refined until the calculated density satisfies the equations self-consistently. Sophisticated software packages are commonly employed to carry out these calculations, automating the process and providing visualization tools to analyze the results. However, a deep understanding of the underlying theory and the limitations of different approximations is essential for interpreting and validating the output. Proper selection of the functional and careful consideration of basis sets greatly influence the reliability of DFT predictions.

The power of DFT lies in its ability to handle complex systems. It enables prediction of a wide range of material properties, including band structures, which determine electrical conductivity, density of states, crucial for understanding optical properties, and magnetic moments, underlying the magnetic behaviour of materials. Moreover, DFT is not limited to static properties. Time-dependent DFT (TDDFT) extends its capabilities to simulate dynamic phenomena, such as the response of materials to light, opening up possibilities for studying optical and spectroscopic properties. DFT-based calculations are routinely used in materials design and discovery, helping researchers to explore and predict the properties of new materials. For example, it has been instrumental in designing novel high-efficiency photovoltaic materials and in guiding the development of advanced catalysts for various chemical reactions.

However, it's critical to understand DFT's limitations. It is a ground-state theory, meaning it primarily provides information about the lowest-energy state of a system. Excited states require more sophisticated techniques, such as TDDFT, but even then, accuracy can be challenging. Furthermore, strong electron correlation, where electrons strongly interact with each other, poses significant difficulties for many commonly used approximations. Systems with significant electron correlation often require more computationally expensive methods, or more advanced functionals. Despite these limitations, DFT remains a cornerstone of modern condensed matter physics and materials science, offering an unparalleled tool for understanding and predicting the properties of a vast range of materials. Its broad applicability, combined with continuous improvements in both approximations and computational capabilities, solidifies its place as an indispensable technique for decades to come. The insights gained from DFT calculations are increasingly influencing the development of advanced materials across various technological sectors.

4 Magnetism

Diamagnetism, the simplest form of magnetism, arises from the orbital motion of electrons. When an external magnetic field is applied, these orbiting electrons experience a Lorentz force, altering their orbits slightly. This induces a small magnetic moment that opposes the applied field, a phenomenon described by Lenz's law. The effect is weak, and its magnitude is proportional to the applied field strength and inversely proportional to the electron mass. It's a fundamental property of all materials, although often masked by stronger magnetic effects. Understanding diamagnetism requires grasping the quantum mechanical nature of electron motion, showcasing the interplay between classical and quantum descriptions highlighted in the book's introduction. The inherent complexity lies in the subtle shifts of electron orbitals, accurately modeled only through quantum mechanics. This seemingly simple phenomenon illustrates the necessity of quantum-mechanical considerations in understanding macroscopic properties. It forms a foundational element for our later discussions on more complex magnetic behaviors.

Paramagnetism, in contrast to diamagnetism, involves a net magnetic moment. Unlike diamagnetism, where the induced moment opposes the applied field, paramagnetic materials exhibit a net magnetic moment aligned with the field. This arises from the presence of unpaired electrons with intrinsic spin angular momentum. Each unpaired electron possesses a magnetic moment, and in the absence of an external field, these moments are randomly oriented. Applying an external field aligns these moments, resulting in a net magnetization parallel to the field. The degree of alignment depends on the temperature and the strength of the applied field, governed by Curie's Law. This describes a simple inverse relationship between the magnetic susceptibility and temperature, demonstrating how thermal fluctuations compete with the ordering influence of the magnetic field. This is a crucial distinction from diamagnetism, which shows negligible temperature dependence. This makes paramagnetism a far more observable phenomenon in many materials at reasonable temperatures.

Ferromagnetism is far more dramatic. It's characterized by spontaneous magnetization, meaning that even without an externally applied field, the material exhibits a significant net magnetization. This arises from a strong coupling between electron spins, called exchange interaction, that leads to a parallel alignment of electron spins over macroscopic regions known as domains. Each domain acts like a tiny magnet, and in the absence of an external field, these domains are randomly oriented, resulting in a zero net magnetization. However, applying an external field aligns these domains, leading to a significant increase in magnetization. The strength of ferromagnetism is substantial, multiple orders of magnitude stronger than paramagnetism, readily evident in common materials like iron, cobalt, and nickel. The energy associated with this spontaneous magnetization is substantial and has significant implications for technological applications. Understanding the underlying exchange interactions requires a deeper dive into quantum mechanics, specifically the Pauli exclusion principle and electron correlation effects. A comprehensive understanding of these interactions is fundamental to this chapter.

Antiferromagnetism, on the other hand, represents another fascinating arrangement of electron spins. Unlike ferromagnetism, where neighboring spins align parallel, in antiferromagnetic materials, neighboring spins align antiparallel. This leads to a zero net magnetization in the absence of an external field. Similar to ferromagnetism, this ordering stems from exchange interactions, but in this case, the interaction favors antiparallel spin alignment. While the net magnetization is zero, there is still a significant magnetic order at the microscopic level, detectable via techniques like neutron scattering. Below a certain temperature (Néel temperature), the antiparallel spin arrangement establishes long-range order. This characteristic is significant for various material properties and their uses. Antiferromagnetism is less prevalent in everyday applications than ferromagnetism, but it plays a pivotal role in fundamental material science and is increasingly relevant in modern technological devices, especially those seeking to harness the unique characteristics of antiferromagnetic materials. This section builds upon our understanding of exchange interactions, highlighting how subtle differences in these interactions lead to dramatically different macroscopic magnetic properties.

The interplay between these different forms of magnetism is complex and often depends on factors such as temperature, applied magnetic field, and crystal structure. For instance, some materials might exhibit paramagnetism at high temperatures but undergo a transition to ferromagnetism or antiferromagnetism at lower temperatures. This transition is characterized by a critical temperature, the Curie temperature for ferromagnetism and the Néel temperature for antiferromagnetism. Furthermore, the crystal structure plays a crucial role in determining the type and strength of magnetic ordering. Different crystal structures can lead to different exchange interactions, ultimately affecting the magnetic behavior of the material. Understanding these factors requires a comprehensive knowledge of both classical and quantum mechanics, reflecting the core theme of this book. This section offers a practical framework for understanding the diverse magnetic behaviors exhibited in condensed matter systems, ranging from simple diamagnetism to complex antiferromagnetism.

4.1 Diamagnetism and Paramagnetism

Diamagnetism arises from the Lenz's law, a fundamental principle in electromagnetism. When an external magnetic field is applied to a material, the electrons within the atoms adjust their orbital motion to generate a small induced magnetic field that opposes the applied field. This induced field is always diamagnetic, meaning it repels the external magnetic field. The effect is quite weak and only observable in materials lacking unpaired electrons, because the diamagnetic response is always present, though often masked by stronger paramagnetic or ferromagnetic effects. It's a fundamental quantum mechanical phenomenon that highlights the intimate connection between electromagnetism and the behavior of electrons at the atomic level. This subtle response is crucial to understanding the overall magnetic behavior of a material. It's a passive response; the material doesn't exhibit magnetism of its own but subtly interacts with external fields. The magnitude of this diamagnetic susceptibility is generally small and negative, often of the order of 105 to 106. This small value reflects the relatively weak nature of the diamagnetic response. It's dependent on the number of electrons in the material and their orbital configuration, showing a direct link between the microscopic structure and macroscopic properties. Understanding diamagnetism gives insight into the fundamental electron behavior and allows better characterization of materials.
Paramagnetism, in contrast to diamagnetism, represents a positive re-

sponse to an external magnetic field. Unlike diamagnetism, which is an induced response based on Lenz's law, paramagnetism stems from the presence of unpaired electrons possessing intrinsic magnetic moments. These unpaired electrons behave like tiny magnets, each aligning its magnetic moment with an applied external field. This alignment enhances the overall magnetic field within the material, leading to a positive susceptibility. The magnitude of the paramagnetic susceptibility varies with temperature, usually following a Curie law: = C/T, where is the susceptibility, C is the Curie constant, and T is the absolute temperature. The Curie law indicates that paramagnetic susceptibility decreases as temperature increases due to thermal agitation, which disrupts the alignment of the magnetic moments. Understanding the Curie Law helps predict the paramagnetic behavior in various materials at different temperatures. The constant C incorporates intrinsic magnetic moment properties of the atoms in the material and is a useful constant for characterization. Thus the temperature dependence is crucial to understanding paramagnetic materials.

The distinction between diamagnetism and paramagnetism is crucial for understanding material properties. While diamagnetism is a universal phenomenon, its effect is generally weak and often overshadowed by paramagnetism in materials with unpaired electrons. Paramagnetism results from the inherent magnetic moments of unpaired electrons and is temperature-dependent, typically showing increased susceptibility at lower temperatures and decreased at higher temperatures. The experimental observation of diamagnetic and paramagnetic responses involves measuring the magnetization of a material in the presence of an external magnetic field. Techniques like SQUID magnetometry allow highly precise measurements of magnetic susceptibility, providing quantitative data for determining whether a material is diamagnetic or paramagnetic. The sign and magnitude of the susceptibility directly indicate the type of magnetism present. This is critical for understanding the interplay between various types of magnetism in materials.

Determining whether a material exhibits diamagnetism or paramagnetism is a significant step in characterizing its magnetic properties. This requires careful measurements of magnetic susceptibility, usually employing highly sensitive techniques like SQUID magnetometry. Analyzing the temperature dependence of the susceptibility also provides crucial insights into the underlying mechanism of magnetism. For instance, the observation of a temperature-independent negative susceptibility points towards diamagnetism. On the other hand, a temperature-dependent positive susceptibility, following the Curie law, strongly suggests paramagnetism. Such experimental data, combined with knowledge of the material's electronic structure, helps confirm and categorize the magnetic response. This is a critical step in classifying a material based on its magnetic properties and understanding its behavior in external magnetic fields. Therefore, practical applications relying on precise magnetic characterization must consider these fundamental magnetic responses.

Many materials exhibit both diamagnetism and paramagnetism simultaneously. The net magnetic response is the sum of these opposing effects. In materials with only paired electrons, the diamagnetic contribution dominates. Conversely, the presence of unpaired electrons often leads to a net paramagnetic response, even though diamagnetism is still present. The interplay between these two effects underscores the complexity of the magnetic behavior exhibited by diverse materials. This highlights the importance of understanding both diamagnetism and paramagnetism to fully comprehend the nature of magnetism in any given substance. Therefore, analyzing the relative strengths of diamagnetic and paramagnetic contributions is essential to accurately assess the magnetic properties of materials. A thorough understanding of these basic forms of magnetism forms the foundation for delving into the more complex phenomena like ferromagnetism and antiferromagnetism.

4.2 Ferromagnetism and Antiferromagnetism

Ferromagnetism arises from a cooperative alignment of electron spins within a material. Unlike diamagnetism or paramagnetism, where magnetic moments are induced by an external field and are relatively weak, ferromagnets exhibit spontaneous magnetization, meaning they possess a net magnetic moment even in the absence of an external field. This spontaneous magnetization is a consequence of exchange interactions, specifically the Heisenberg exchange interaction, which favors parallel spin alignment between neighboring atoms. The strength of this interaction is crucial in determining the Curie temperature, the temperature above which thermal fluctuations overcome the exchange interaction, leading to a loss of spontaneous magnetization and a transition to a paramagnetic state. The alignment of spins creates magnetic domains, regions of uniformly aligned spins, which can be manipulated by external magnetic fields. Understanding the size, shape and dynamics of these domains is key to understanding the macroscopic magnetic properties of ferromagnetic materials, their technological applications ranging from hard disk drives to MRI machines. The detailed microscopic analysis is often accomplished using techniques like neutron scattering.

The energy associated with the Heisenberg exchange interaction is a direct consequence of the Pauli exclusion principle and the Coulomb interaction between electrons. Electrons with parallel spins tend to avoid each other more effectively, reducing their Coulomb repulsion, which leads to a lower total energy compared to the antiparallel configuration. This energetic preference for parallel spins is the driving force behind ferromagnetism. However, the specifics of the exchange interaction are often material-dependent and can be quite complex, especially in systems with strong spin-orbit coupling or multiple magnetic ions. Theoretical modeling of these systems often requires advanced computational techniques beyond simple mean-field approximations. Experimental techniques, such as magnetometry and magnetic resonance, provide valuable insights into the magnetic properties of ferromagnets, but often require careful interpretation to disentangle the contributions from different microscopic mechanisms.

Antiferromagnetism represents another fascinating manifestation of magnetic ordering in condensed matter systems. In contrast to ferromagnets, antiferromagnets exhibit an antiparallel alignment of neighboring spins, resulting in a net magnetization of zero in the absence of an external magnetic field. While there is no net magnetic moment on a macroscopic scale, antiferromagnets still exhibit strong magnetic correlations on a microscopic level. These correlations influence various physical properties, including electrical resistivity, thermal conductivity, and optical response. The Néel temperature, analogous to the Curie temperature in ferromagnets, marks the transition from the antiferromagnetically ordered state to a paramagnetic state. Below the Néel temperature, the antiparallel alignment of spins persists, creating a complex magnetic structure. The determination of this structure often requires sophisticated experimental techniques such as neutron diffraction, which probes the magnetic structure directly through the interaction of neutrons with atomic spins.

The exchange interaction is once again a key player in antiferromagnetism, but in this case, it favors antiparallel spin alignment. This preference often arises from superexchange, where magnetic ions are not directly bonded but interact indirectly through intervening non-magnetic ions. The specific arrangement of ions and the nature of the chemical bonds are critical in determining the type of magnetic ordering. The presence of competing interactions can lead to more complex magnetic structures, such as ferrimagnetism, where unequal numbers of spins aligned parallel and antiparallel result in a nonzero net magnetization. Furthermore, the application of an external magnetic field can significantly influence the magnetic properties of antiferromagnets, leading to phenomena such as spin-flop transitions, where the antiparallel spin alignment reorients under the influence of the field. These transitions highlight the subtle interplay between exchange interactions and external fields, offering profound insights into the fundamental physics of magnetic systems.

Understanding ferromagnetism and antiferromagnetism is not merely an academic exercise. These phenomena are fundamental to the operation of numerous technological devices. Ferromagnetic materials are essential components in hard disk drives, transformers, and electric motors, and their magnetic properties are carefully engineered to optimize performance. Antiferromagnetic materials, while not directly exploited for their magnetization, are increasingly recognized for their potential applications in spintronics, a field that aims to manipulate electron spins for information processing and storage. The ability to control and manipulate spin currents in antiferromagnetic materials opens up possibilities for faster and more energy-efficient devices. Recent research has even explored the use of antiferromagnets for designing novel memory devices with potential advantages over conventional ferromagnetic-based technologies, leading to improvements in data storage density and speed. The ongoing research on magnetic materials and systems promises exciting new advancements in various technological fields.

5 Superconductivity

Superconductivity is a remarkable phenomenon where certain materials exhibit zero electrical resistance below a critical temperature. This complete absence of resistance allows for the flow of electric current without any energy loss, a property with immense technological potential. Imagine power transmission lines with no energy dissipation, or incredibly efficient energy storage systems; these are just a few examples of the transformative impact superconductivity could have. However, understanding the underlying physics is key to unlocking its full potential.

The Meissner effect is a defining characteristic of superconductors. When a material transitions into the superconducting state, it expels magnetic fields from its interior. This expulsion isn't simply a passive blocking of magnetic flux; it's an active process where the material generates currents that precisely cancel out any applied magnetic field. This macroscopic manifestation of quantum mechanics is a testament to the extraordinary behavior found in condensed matter systems. Observing the Meissner effect provides a clear, experimental indication that a material has become superconducting.

The microscopic explanation for superconductivity in conventional materials—those that become superconducting at relatively low temperatures—is provided by the Bardeen-Cooper-Schrieffer (BCS) theory. This theory postulates that electrons in a superconductor form bound pairs, known as Cooper pairs, due to an indirect attractive interaction mediated by lattice vibrations (phonons). These phonons, quantized lattice vibrations, create a slight distortion in the crystal lattice, affecting the potential energy experienced by other electrons. This indirect interaction overcomes the Coulomb repulsion between electrons, leading to the formation of Cooper pairs. The formation of these pairs is crucial; it allows the electrons to move collectively, avoiding scattering from the lattice and hence, eliminating electrical resistance.

However, the BCS theory, while highly successful in explaining low-temperature superconductivity, doesn't fully account for the behavior of high-temperature superconductors (high-Tc superconductors). These materials exhibit superconductivity at significantly higher temperatures than predicted by the BCS theory, often above the temperature of liquid nitrogen, making them much more practical for many applications. The exact mechanism driving high-Tc superconductivity remains a topic of intense research and debate. Several theories exist, but none offer a complete and universally accepted explanation. This ongoing research highlights the continuing need for further exploration and discovery in condensed matter physics. Understanding high-Tc superconductors could revolutionize energy technologies and countless other applications.

The technological implications of superconductivity are vast and far-reaching. The potential applications extend beyond the aforementioned power transmission and energy storage. Superconducting magnets are already used in magnetic resonance imaging (MRI) machines, providing high-resolution medical scans. They are also crucial components in particle accelerators, pushing the boundaries of our understanding of fundamental physics. Furthermore, researchers are actively exploring the use of superconductors in various electronic devices,

aiming to create significantly faster and more energy-efficient computing technologies. The possibility of lossless energy transport alone promises a dramatic reduction in energy consumption, benefiting both the environment and the global economy.
Challenges remain, of course. The need for extremely low temperatures for many superconductors limits their widespread use. While high-Tc superconductors offer a potential solution, these materials often have complex crystal structures and are difficult to synthesize in a high-quality form suitable for technological applications. Furthermore, understanding and improving the critical current density, a measure of how much current a superconductor can carry before losing its superconducting properties, remains a key area of research. The development of new materials with improved properties and the advancement of existing technologies are essential steps toward realizing the full potential of superconductivity. The quest for room-temperature superconductors, a long-standing goal in the field, continues to drive innovation and exploration. The discovery of such a material would indeed be a scientific and technological breakthrough of unparalleled significance. But even without achieving room temperature superconductivity, the ongoing progress in this area continues to yield transformative advancements in various fields.

5.1 Meissner Effect

The Meissner effect, a cornerstone of superconductivity, demonstrates a material's remarkable ability to expel magnetic fields from its interior once it transitions below its critical temperature. Imagine a magnet hovering effortlessly above a superconducting disk; this seemingly magical levitation is a direct consequence of the Meissner effect. It's not simply the absence of resistance; superconductors actively repel magnetic flux, a truly unique characteristic. This expulsion isn't a passive phenomenon, a mere consequence of zero resistance; instead, it's an active process, a fundamental property intrinsic to the superconducting state. This active expulsion, crucial for understanding superconductivity's applications, is the key to many technological advancements. The effect arises from the interplay between the superconductor's electrons and an external magnetic field, a delicate dance governed by quantum mechanics.

Unlike a perfect conductor, which would simply prevent magnetic flux from penetrating after it's established, a superconductor actively expels any existing magnetic field once it enters the superconducting state. This difference is profound and is the defining characteristic distinguishing a perfect conductor from a superconductor. The magnetic field lines are not merely blocked, they are actively pushed out, creating a region of zero magnetic field within the superconductor. This complete expulsion of magnetic flux, a strikingly different behavior from classical conductors, represents a powerful manifestation of quantum phenomena in macroscopic systems. Understanding this behavior is essential for the design and implementation of various superconducting devices.

Let's delve into the microscopic explanation. The Meissner effect stems from the formation of Cooper pairs, pairs of electrons bound together by an attractive interaction mediated by lattice vibrations (phonons). These pairs behave as bosons, unlike individual electrons which are fermions. This bosonic nature allows them to condense into a macroscopic quantum state, a coherent superposition spanning the entire superconductor. This condensate then generates screening currents that exactly cancel the applied magnetic field within the material. This process is not merely a consequence of the material's properties, but it arises from the cooperative behavior of the many electrons in the superconducting state. The collective quantum nature of these electrons, acting as a single entity, is responsible for this remarkable effect. The energy cost of allowing a magnetic field inside the superconductor becomes prohibitively large, resulting in its expulsion.

It's crucial to understand that the Meissner effect is not merely a consequence of perfect conductivity. A perfect conductor, according to classical electromagnetism, would prevent a change in magnetic flux but wouldn't expel an already existing magnetic field. The Meissner effect, however, does precisely that, demonstrating a crucial difference between the two concepts. This distinction underscores the profoundly quantum mechanical nature of superconductivity, highlighting the limitations of classical descriptions in capturing its essence. The expulsion of the magnetic field is an active process, not a passive consequence of zero electrical resistance. It's a manifestation of macroscopic quantum phenomena.

The practical implications of the Meissner effect are enormous. The perfect diamagnetism exhibited by superconductors is exploited in numerous applications, from powerful superconducting magnets used in MRI machines to levitating trains utilizing the magnetic repulsion for frictionless transportation. Further research into novel superconductors, particularly those operating at higher temperatures, promises even more transformative technologies. The quest for room-temperature superconductors remains a significant challenge in condensed matter physics, but the potential rewards—efficient energy transmission, powerful and compact electronics—are immense. Understanding the Meissner effect, therefore, is not merely an academic exercise, but it is the key to unlocking a wide range of technological breakthroughs in numerous fields.

Furthermore, studying the Meissner effect provides crucial insights into the fundamental physics governing superconductivity. It serves as a powerful experimental tool for probing the nature of the superconducting state and testing theoretical models. Detailed measurements of the magnetic field expulsion can reveal information about the penetration depth, a characteristic length scale defining how deeply the magnetic field can penetrate the superconductor before being expelled. This depth is directly related to the density of the superconducting electron pairs, providing valuable information about the superconducting condensate. The Meissner effect is thus not only a technological marvel but also a fundamental tool for furthering our understanding of this remarkable phenomenon. Its study continues to reveal the intricate details of the quantum world's influence on macroscopic matter. This effect continues to challenge our understanding of matter and inspire new research avenues. The ongoing research promises new innovations.

5.2 BCS Theory

The Bardeen-Cooper-Schrieffer (BCS) theory, a cornerstone of condensed matter physics, elegantly explains conventional superconductivity. It posits that electrons, normally repelling each other due to their like charges, can form bound pairs, known as Cooper pairs, under specific conditions. This pairing arises from an indirect attractive interaction mediated by the lattice vibrations – the phonons. Imagine the lattice as a trampoline; as an electron moves through it, it deforms the lattice, creating a slight positive charge region. A second electron, feeling this attraction, is drawn towards the first, overcoming their natural electrostatic repulsion. This seemingly counterintuitive effect is crucial to understanding the phenomenon.

The formation of Cooper pairs is not a simple matter of direct attraction. Instead, it involves a complex interplay between the electron's interaction with the lattice and the resulting distortion. This indirect interaction is relatively weak, but at low enough temperatures it becomes significant enough to overcome the electron's inherent Coulomb repulsion. The resulting pairs behave quite differently from individual electrons. They have a net zero momentum and possess a characteristic energy gap – an energy barrier preventing the excitation of electrons, which directly leads to the loss of electrical resistance. The lower the temperature, the stronger this pairing becomes, creating a more robust superconducting state.

To truly grasp the BCS theory, one must delve into its mathematical framework. This involves the use of quantum field theory techniques to describe the interactions between electrons and phonons. The theory leads to a set of equations that predict the critical temperature (Tc), below which superconductivity emerges. This temperature is a key parameter; it signifies the transition point where the attractive interaction between Cooper pairs overwhelms the disruptive thermal energy. The BCS theory provides a remarkably accurate description of the critical temperature for many conventional superconductors, showing a clear correlation between Tc and material properties like phonon frequencies and electron density of states.

The success of the BCS theory lies not just in its predictions of Tc, but also in its explanation of other key superconducting properties. The Meissner effect, the complete expulsion of magnetic fields from the superconductor, is a direct consequence of the formation of Cooper pairs and their collective behavior. These pairs act as a single entity, and their coordinated motion prevents magnetic field penetration. Moreover, the theory successfully explains the energy gap observed in superconductors, a characteristic feature observable through various experimental techniques like tunneling spectroscopy. These experimental verifications provide compelling evidence supporting the core principles of the theory.

However, the BCS theory is not a universal explanation of superconductivity. Its applicability is limited to conventional superconductors, materials exhibiting relatively low critical temperatures. The discovery of high-Tc superconductors, materials with significantly higher critical temperatures, posed a considerable challenge to the theory. These materials often exhibit properties that deviate significantly from the predictions of the BCS theory, suggesting the involvement of other mechanisms beyond the simple electron-phonon interaction. Despite this limitation, the BCS theory remains a monumental achievement in condensed matter physics, providing a fundamental framework for understanding conventional superconductivity and serving as a springboard for the exploration of more exotic superconducting phenomena.

The mathematical formalism of the BCS theory involves several complex steps, often requiring a strong background in quantum mechanics and many-body physics. However, the underlying concepts are conceptually accessible. Understanding the phonon-mediated attraction between electrons is key to grasping the formation of Cooper pairs. Visualizing this process helps in comprehending the fundamental reason behind the superconducting state's emergence. The crucial role of temperature in determining the balance between attractive pairing and thermal disruptions is another essential aspect to remember.

Furthermore, the importance of the energy gap cannot be overstated. This energy barrier prevents the scattering of electrons, leading to the absence of electrical resistance. The relationship between the energy gap and the critical temperature, as predicted by the BCS theory, is a powerful tool for understanding and characterizing superconducting materials. This concept is particularly useful in relating the macroscopic properties (like the critical temperature) to the microscopic interactions within the material.

The BCS theory provides a powerful tool for understanding superconductivity and lays a strong foundation for future research. It explains the basic mechanism behind the phenomenon in many materials, leading to numerous practical applications. The theory's limitations with high-Tc superconductors spur ongoing research into more complex interactions and mechanisms responsible for their superconducting properties. The legacy of BCS theory extends beyond its direct applications; it exemplifies the power of theoretical physics in explaining complex phenomena and inspiring further investigation into the intricacies of the quantum world. It reminds us that even seemingly simple interactions at the atomic level can lead to remarkable macroscopic effects, driving innovations and discoveries that continue to shape our technological landscape.

5.3 High-Tc Superconductors

High-temperature superconductivity remains one of the most captivating and challenging areas in condensed matter physics. Unlike conventional superconductors, which are explained by the Bardeen-Cooper-Schrieffer (BCS) theory involving electron-phonon interactions, high-Tc superconductors exhibit superconductivity at significantly higher temperatures, often exceeding the boiling point of liquid nitrogen. This remarkable behavior defies simple BCS explanations, necessitating the exploration of more complex mechanisms. The discovery of these materials, specifically cuprates like YBCO (Yttrium Barium Copper Oxide), revolutionized the field and continues to spark intense research.

The defining characteristic of high-Tc superconductors is their unexpectedly high critical temperatures (Tc), far exceeding the predictions of the BCS theory. These materials often involve layered structures, exhibiting anisotropic properties; their superconducting properties can vary significantly depending on the direction of the current or magnetic field. This anisotropy is deeply connected to the complex electronic structure and the specific arrangement of atoms within the crystal lattice. Understanding this structure-property relationship is paramount to developing a complete theoretical understanding.

Many high-Tc superconductors are copper-oxide based, featuring copper-oxygen planes interspersed with other layers. These copper-oxygen planes are believed to be crucial for the superconducting mechanism. The electrons within these planes are thought to interact strongly, exhibiting a behavior that deviates significantly from the independent-electron picture assumed in the simpler BCS model. It's within this interplay of strong electron correlations and the layered structure that the key to high-Tc superconductivity likely lies.

While a universally accepted theory for high-Tc superconductivity remains elusive, several competing models attempt to explain the phenomenon. Some models emphasize the role of strong electron-electron correlations, suggesting that the superconducting state arises from the pairing of electrons mediated by fluctuations rather than phonons. These fluctuations, often antiferromagnetic in nature, can lead to a pairing mechanism distinct from the phonon-mediated pairing in conventional superconductors.

Other models focus on the layered structure of the materials, proposing that the interaction between the copper-oxygen planes plays a crucial role in the emergence of high-Tc superconductivity. These models often involve the concept of "interlayer coupling," where interactions between adjacent layers influence the electronic properties of the entire material. This complexity highlights the intricate interplay between structure and electronic behavior. The intricate details of the electronic band structure, often investigated through techniques like angle-resolved photoemission spectroscopy (ARPES), provide crucial insights into these interactions.

The experimental investigation of high-Tc superconductors involves a wide range of techniques. Beyond ARPES, researchers utilize techniques like neutron scattering to study the magnetic properties and phonon spectra. Scanning tunneling microscopy (STM) provides real-space images of the electronic structure, revealing nanoscale variations and inhomogeneities that can significantly influence the superconducting properties. These experimental observations are essential in validating theoretical models and guiding further research.

The technological potential of high-Tc superconductors is enormous. Their ability to operate at higher temperatures compared to conventional superconductors makes them potentially suitable for a wide range of applications, including power transmission, magnetic resonance imaging (MRI) machines, and high-speed electronic devices. However, significant challenges remain. Many high-Tc superconductors are brittle and difficult to process into useful forms. Improving their material properties, including mechanical strength and current carrying capacity, is a crucial area of ongoing research and development.

Understanding high-temperature superconductivity isn't merely an academic pursuit; it has profound technological implications. The development of practical high-Tc superconducting wires and devices could revolutionize various industries, improving energy efficiency and leading to the development of novel technologies. However, the road to realizing this potential is long and challenging, requiring a multidisciplinary approach that combines materials science, condensed matter physics, and engineering. The journey toward a comprehensive understanding continues to attract researchers globally, driven by both scientific curiosity and the promise of technological breakthroughs. Continued exploration and innovation in this field hold the key to unlocking the full potential of this extraordinary phenomenon. Further research will need to focus on refining existing theoretical models and developing new experimental techniques to uncover the underlying secrets of high-Tc superconductivity. Only through a combined approach can we unlock the full potential of these materials and translate them into real-world applications. This remains one of the grand challenges of condensed matter physics, promising a future rich in both scientific understanding and technological advancement.

6 Quantum Hall Effect

The quantum Hall effect, a striking manifestation of quantum mechanics in condensed matter systems, arises when a two-dimensional electron gas (2DEG) is subjected to a strong perpendicular magnetic field and low temperatures. Under these conditions, the electrons' motion becomes quantized, leading to the formation of Landau levels—discrete energy levels separated by energy gaps. This quantization isn't just a theoretical curiosity; it has profound experimental consequences, most notably the appearance of plateaus in the Hall resistance. These plateaus are remarkably precise and independent of the details of the sample's disorder, a characteristic that initially baffled researchers but ultimately paved the way for the development of high-precision resistance standards.

The integer quantum Hall effect (IQHE), the first discovered of these phenomena, exhibits plateaus in the Hall resistance at values of RH = h/e2, where h is Planck's constant, e is the elementary charge, and is an integer. This quantized resistance, directly proportional to the fundamental constants of nature, is a testament to the power of quantum mechanics. The observation of these plateaus is a direct consequence of the formation of edge states, chiral one-dimensional channels carrying current along the sample's edges. These edge states are robust against disorder and are responsible for the remarkable precision of the quantized Hall resistance.

Moving beyond the integer case, the fractional quantum Hall effect (FQHE) reveals even more surprising behavior. Observed at lower temperatures and stronger magnetic fields, the FQHE displays Hall resistance plateaus at fractions of h/e2, such as 1/3, 2/5, and 3/7. This fractional quantization is a clear indication that the electrons are interacting strongly with one another, forming novel collective states. The theoretical explanation, involving the formation of composite fermions—quasiparticles consisting of electrons bound to magnetic flux quanta—provided a breakthrough in our understanding of strongly correlated electron systems. These composite fermions behave like non-interacting particles in a modified magnetic field, effectively mapping the complicated many-body problem onto a simpler one.

Furthermore, the FQHE is intimately connected to the concept of topology. The quantized Hall conductance is a topological invariant, meaning that it's robust against smooth changes in the system's parameters, like small amounts of disorder or changes in the electron density. This topological character is responsible for the remarkable stability of the FQHE plateaus. It's crucial to understand that this topological nature isn't simply a mathematical curiosity; it has profound implications for the stability and resilience of these quantum states. The robustness against perturbations makes the FQHE a powerful tool for building fault-tolerant quantum computing devices.

The relationship between the quantum Hall effect and topological insulators underscores the deep connections between these seemingly disparate phenomena. Topological insulators are materials that are insulating in their bulk but possess conducting edge or surface states protected by topology. These edge states are similar in nature to those found in the quantum Hall effect, exhibiting a unique spin-momentum locking, meaning that the electrons' spin is locked to their momentum direction. This inherent spin-momentum locking makes these edge states exceptionally resistant to backscattering, leading to dissipationless transport along the edges. The discovery and exploration of these novel materials have opened up exciting possibilities for spintronics and quantum computing, offering promising avenues for building devices based on topological protection. The fundamental understanding of topological insulators requires a deep understanding of the concepts underlying the quantum Hall effect, building a bridge between seemingly disparate areas of condensed matter physics.

Finally, the quantum Hall effect's experimental realization has been instrumental in advancing our understanding of fundamental physics and developing novel technologies. The extreme precision of the quantized Hall resistance has led to its adoption as a fundamental standard for electrical resistance measurements, underscoring its practical implications. Beyond metrology, the study of the quantum Hall effect continues to provide valuable insights into the behavior of strongly correlated electron systems and the intriguing world of topological matter. Ongoing research continues to explore the ramifications of this remarkable phenomenon, pushing the boundaries of our knowledge and paving the way for future technological advancements. From fundamental physics to cutting-edge technology, the quantum Hall effect serves as a testament to the power and beauty of quantum phenomena in condensed matter systems.

6.1 Integer Quantum Hall Effect

The Integer Quantum Hall Effect (IQHE) is a stunning example of how quantum mechanics manifests itself in condensed matter systems. Imagine a two-dimensional electron gas, a thin layer of electrons confined to a plane, subjected to a strong perpendicular magnetic field.

Classically, one might expect a continuous increase in the Hall resistance (the voltage transverse to the current flow) as the magnetic field strength increases. However, the quantum world offers a dramatically different picture. At low temperatures, the Hall resistance doesn't increase smoothly; instead, it plateaus at quantized values, precisely integer multiples of h/e^2, where h is Planck's constant and e is the elementary charge. These plateaus are remarkably precise, exhibiting an accuracy that rivals the best standards in metrology. This quantization is not merely an approximation; it's a fundamental consequence of the electrons' wave-like behavior in the presence of a strong magnetic field. The key to understanding this remarkable phenomenon lies in the formation of Landau levels. The magnetic field confines the electrons to cyclotron orbits, their energies becoming quantized into discrete levels. Crucially, these levels are highly degenerate, meaning many electrons can occupy the same energy level. The filling of these Landau levels dictates the Hall resistance. When a Landau level is completely filled, the Hall resistance plateaus at an integer multiple of h/e^2. Imperfections in the sample, such as impurities or edge states, determine the precise location of the plateaus, while the quantization itself is a robust consequence of fundamental quantum mechanics.

The precise quantization of the Hall resistance has profound implications. Firstly, it provides an extremely accurate way to measure fundamental constants like Planck's constant and the elementary charge. The accuracy achieved through the IQHE has made it a cornerstone of modern metrology, enabling exceptionally precise measurements that underpin various technological applications. Furthermore, the discovery of the IQHE stimulated significant theoretical advances in our understanding of topological phases of matter. The quantized Hall resistance is linked to a topological invariant, a characteristic that remains unchanged under smooth deformations of the system. This topological nature implies a degree of robustness, making the quantized value insensitive to small changes in the system's parameters. This topological robustness is a hallmark of many other intriguing quantum phenomena, driving ongoing research in topological insulators and superconductors.

Consider the experimental setup: a high-mobility two-dimensional electron gas, typically created in a semiconductor heterostructure, is cooled to extremely low temperatures, often close to absolute zero. A strong magnetic field is then applied perpendicular to the plane of the electron gas. As the magnetic field is increased, the Hall resistance is measured using a four-probe technique, carefully eliminating the influence of longitudinal resistance. The observed plateaus in the Hall resistance provide direct evidence of the IQHE. The experimental precision is key: the observed quantization is so accurate it surpasses the capabilities of conventional measurement methods. The remarkable stability of the Hall resistance plateaus makes the IQHE a compelling example of a macroscopic quantum phenomenon, where quantum effects are observable on a large scale.

The theoretical understanding of the IQHE has been a major triumph of theoretical condensed matter physics. The combination of the Landau levels, the role of edge states, and the connection to topology provides a cohesive explanation for the observed quantization. While edge states are crucial for current flow in this two-dimensional system, they do not contribute directly to the quantization of the Hall resistance. The bulk states, filling the Landau levels, and the topological nature of the system are responsible for the precise quantization observed experimentally. Furthermore, the IQHE offers a unique platform for exploring fundamental aspects of quantum mechanics, such as the quantized nature of charge and the concept of topological order. The theoretical framework surrounding the IQHE has inspired and guided research on other quantum phenomena, highlighting the central role of topology in understanding the behavior of condensed matter systems. The IQHE continues to be an area of active research, prompting new discoveries and offering novel perspectives on the fundamental laws governing the quantum world. The enduring fascination with the IQHE stems from its remarkable precision, its profound theoretical implications, and its role as a stepping stone to understanding more complex quantum phenomena. Its impact extends far beyond basic research, significantly influencing metrology and technological advancements, cementing its status as a pivotal discovery in condensed matter physics.

6.2 Fractional Quantum Hall Effect

The fractional quantum Hall effect (FQHE), a mesmerizing phenomenon occurring in two-dimensional electron systems at extremely low temperatures and high magnetic fields, presents a profound departure from the integer quantum Hall effect (IQHE). Unlike the IQHE, where the Hall conductance is an integer multiple of e^2/h (e being the electron charge and h Planck's constant), the FQHE exhibits plateaus at fractional multiples of this fundamental quantum of conductance. This observation, initially perplexing, points to the formation of novel quasiparticles with fractional electric charge, a striking manifestation of the many-body interactions governing the electrons' behavior. These quasiparticles, known as composite fermions, are not fundamental particles but rather emergent entities resulting from complex correlations between the electrons and the magnetic field. Their existence challenges our classical intuition and provides a fertile ground for exploring the intricacies of strongly correlated systems.
Understanding the FQHE demands a move beyond the independent electron picture that successfully describes the IQHE. Here, the strong

Coulomb repulsion between electrons cannot be ignored. Instead of independent electrons moving in a uniform magnetic field, we must consider the complex interplay of their mutual interactions within the confines of the two-dimensional electron gas (2DEG). These interactions lead to the formation of an incompressible quantum fluid, a state of matter characterized by a substantial energy gap separating the ground state from excited states. This energy gap is responsible for the observed quantization of the Hall conductance, making it exceptionally robust against small perturbations. The formation of this incompressible state is a testament to the power of quantum correlations in dictating the macroscopic behavior of the system.

The concept of composite fermions offers a particularly illuminating framework for understanding the FQHE. These quasiparticles are essentially electrons bound to an even number of magnetic flux quanta. This binding effectively reduces the magnetic field experienced by the composite fermions, allowing them to behave, to a good approximation, as independent particles in a weaker effective magnetic field. This effective field can then be tuned by adjusting the filling fraction, which in turn determines the observed Hall conductance. This elegantly explains the emergence of fractional Hall plateaus and the associated fractional charges of the composite fermions. The remarkable success of this composite fermion picture underscores the potential for novel theoretical constructs to unveil the hidden complexity of interacting quantum many-body systems.

Experimental verification of the FQHE is not merely a matter of observing fractional Hall conductance plateaus. The fractional charge of the quasiparticles has been directly measured through shot noise experiments, further solidifying the reality of these emergent entities. These experiments reveal the characteristic noise signature associated with fractional charges, confirming the theoretical predictions based on the composite fermion picture. The observation of fractional charge, a concept seemingly at odds with our fundamental understanding of charge quantization, underscores the remarkable richness of the quantum world. The existence of quasiparticles with fractional charge is a stark reminder that the quantum world often surpasses our classical intuition, leading to unexpected and profound phenomena.

Further investigation of the FQHE reveals its deep connection to topology. The ground states associated with the different fractional Hall plateaus possess distinct topological properties. This topological order is responsible for the robustness of the FQHE against perturbations, as even significant changes to the system's microscopic details may not alter the topological properties of the ground state. This deep connection to topology has not only provided a more fundamental understanding of the FQHE but also spawned the field of topological quantum computing, where robust quantum information storage and manipulation are sought using the topological properties of materials. The FQHE, therefore, transcends its initial position as a curious quantum effect, becoming a cornerstone for advancements in quantum information science.

Beyond its fundamental importance, the FQHE serves as a powerful model system for studying more complex strongly correlated electron systems. The theoretical tools developed to understand the FQHE, such as Chern-Simons theory and the composite fermion approach, provide valuable insights into the behavior of other materials exhibiting strong electron-electron interactions. This includes high-temperature superconductors, where complex interplay between lattice vibrations and electronic correlations plays a critical role in the phenomenon of superconductivity. By studying the FQHE, we gain crucial insights into the techniques needed to decipher the intricate behavior of interacting many-body systems, which remain a significant challenge in condensed matter physics. The principles of the FQHE provide a unique perspective on understanding highly correlated systems across various condensed matter realms. The exploration of such phenomena continuously expands our comprehension of these incredibly complex systems. The journey into the depths of the FQHE continues to uncover fascinating features, pushing the boundaries of our understanding of fundamental physics.

6.3 Topological Insulators

Topological insulators represent a fascinating intersection of quantum mechanics and materials science. Unlike ordinary insulators, where the energy gap between the valence and conduction bands is a simple consequence of band structure, topological insulators possess a unique bulk energy gap but also exhibit conducting states on their surfaces or edges. This seemingly paradoxical behavior stems from the topology of the electronic wave functions, a concept rooted in advanced mathematical descriptions of the system's overall structure and symmetries rather than just its local properties. These surface states are protected by time-reversal symmetry and are exceptionally robust against perturbations or defects, unlike typical metallic surface states that are often easily disrupted. Their existence is a manifestation of the profound link between the material's bulk properties and the boundary conditions.
The defining characteristic of a topological insulator lies in its non-trivial topological invariant, a quantity that cannot change continuously under smooth deformations of the system's Hamiltonian. This invariant provides a fingerprint, if you will, for identifying a topological phase. Imagine it like a rubber band: you can stretch and twist it, but you can't smoothly turn it into a circle without cutting it. Similarly, you can modify a topological insulator in many ways, but its topo-

logical properties, and therefore its surface states, remain protected as long as certain symmetries are preserved. This resilience is crucial for potential applications in spintronics and quantum computing.

One of the most intriguing aspects of topological insulators is the spin-momentum locking of their surface states. This means that the direction of the electron's spin is inextricably linked to its momentum—electrons moving in one direction will have their spins aligned in one direction, while electrons moving in the opposite direction will have spins aligned oppositely. This phenomenon has enormous implications for spintronics—the field that aims to exploit electron spin rather than just its charge to develop new devices. By manipulating the spin orientation of the electrons, one could potentially create much more energy-efficient and faster electronics.

Experimentally, observing these unique properties can be quite demanding. Early research focused primarily on identifying materials that exhibit the hallmarks of topological behavior. This involved advanced characterization techniques, such as angle-resolved photoemission spectroscopy (ARPES), which allows researchers to directly probe the electronic band structure and identify the surface states. Transport measurements are also critical, providing evidence of the conductive surface states while the bulk remains insulating. These experiments require highly sophisticated equipment and meticulous sample preparation to avoid confounding factors and ensure reliable results.

The theoretical understanding of topological insulators draws heavily from concepts like Berry curvature and Chern numbers, which are advanced topological concepts that characterize the electronic states. These mathematical tools go beyond conventional band theory by capturing the global properties of the system. While a detailed mathematical treatment is beyond the scope of this introductory chapter, understanding the basic idea of a non-trivial topological invariant is

key to grasping the nature of these materials. The surface states, and their associated conductivity, are not just an accident but a consequence of the fundamental topological properties of the bulk material itself.
Furthermore, the field of topological insulators continues to be an active area of research with many promising areas of exploration. The search for new materials with improved properties, such as higher operating temperatures or stronger surface conductivity, is ongoing. Moreover, exploring other topological phases, like topological superconductors, which combine superconductivity with topological features, opens entirely new avenues of research. One key challenge is to improve the quality and control of topological insulator materials, which is vital for their potential application in practical devices. The development of new fabrication techniques and improved sample preparation methods will be crucial in realizing the full potential of topological insulators.
The potential applications of topological insulators extend beyond spintronics. Their unique properties are also being explored for their potential use in quantum computation. The highly robust surface states, protected by topology, are relatively insensitive to external disturbances, making them ideal candidates for building stable qubits. This area is still in its nascent stages, however, and much research is needed to overcome the significant challenges in harnessing these properties for functional quantum computing devices.
In summary, topological insulators are an exceptional example of the fascinating interplay between classical and quantum phenomena in condensed matter systems. Their unique combination of bulk insulating behavior and surface conductivity, stemming from intricate topological properties of their electronic structure, holds significant promise for technological applications, particularly in spintronics and quantum computing. However, ongoing research is vital to fully understand and exploit the full potential of these remarkable materials. The explo-

ration of topological insulators underscores the importance of combining experimental investigation with advanced theoretical frameworks to uncover the deeper mysteries of quantum condensed matter physics.

7 Advanced Topics

Quantum computing leverages the principles of quantum mechanics to perform computations in ways that are impossible for classical computers. Solid-state systems, particularly those based on superconducting circuits or trapped ions within crystalline structures, offer promising platforms for building these quantum computers. The manipulation of quantum states – superposition and entanglement – within these carefully engineered materials forms the basis for quantum algorithms capable of solving problems currently intractable to classical computation. For example, Shor's algorithm, implemented using a superconducting qubit array, promises to revolutionize cryptography by efficiently factoring large numbers. The challenge lies in scaling up these systems, maintaining the delicate coherence of quantum states over extended periods, and developing error correction techniques to mitigate the effects of noise and decoherence. Significant progress continues in this field, pushing the boundaries of what's computationally possible and opening doors to new scientific discoveries. Ongoing research focuses on improving qubit fidelity, developing new qubit architectures, and exploring novel materials with improved quantum properties. The future likely holds a hybrid approach, integrating classical and quantum computing resources for optimal performance.

Strongly correlated electron systems exhibit fascinating behavior that cannot be easily understood using independent-electron approximations. In these materials, the interactions between electrons are so strong that they profoundly alter the electronic structure and the resulting macroscopic properties. Examples include high-temperature superconductors, heavy fermion materials, and colossal magnetoresistive oxides. Understanding the intricate interplay of electron-electron interactions, lattice vibrations, and spin degrees of freedom in these systems remains a significant challenge in condensed matter physics. Theoretical approaches, such as dynamical mean-field theory and density matrix renormalization group methods, are employed to tackle the complexity of these systems. Experimental techniques, such as angle-resolved photoemission spectroscopy and neutron scattering, provide crucial insights into their electronic and magnetic structures. The goal is to develop a comprehensive theoretical framework that can accurately predict and explain the diverse and often unexpected properties of strongly correlated materials. This would not only enhance our fundamental understanding of quantum mechanics in condensed matter but also lead to the discovery of novel materials with potentially revolutionary technological applications.

The future of condensed matter physics is bright, brimming with exciting possibilities and unanswered questions. The field continues to advance at a rapid pace, driven by both theoretical breakthroughs and technological innovations. One major focus is the exploration of novel materials with unique properties, including topological insulators and Weyl semimetals, which exhibit exotic electronic band structures and potential applications in spintronics and quantum computing. The development of advanced experimental techniques, such as ultrafast spectroscopy and scanning probe microscopy, provides ever-finer control and observation of materials at the atomic scale. Theoretical efforts are directed towards developing more accurate and efficient computational methods for predicting the properties of complex materials. Furthermore, interdisciplinary collaborations between condensed matter physicists, chemists, material scientists, and engineers are crucial for pushing the boundaries of the field and translating fundamental discoveries into real-world applications. The quest to understand and control quantum phenomena in condensed matter systems promises a future filled with both fundamental scientific advancements and transformative technological innovations. Unraveling the mysteries of quantum materials may hold the key to new paradigms in computing, energy, and materials science, fostering a technological revolution built upon a deeper understanding of the quantum world.

7.1 Quantum Computing with Solids

Harnessing the quantum properties of solids opens up exciting possibilities for quantum computation. Unlike classical bits representing 0 or 1, quantum bits, or qubits, leverage superposition, existing in both states simultaneously. This allows for vastly increased computational power, potentially solving problems intractable for even the most powerful classical computers. Several solid-state systems are promising candidates for building qubits.

Superconducting circuits, for instance, exploit the unique properties of superconductors—materials that exhibit zero electrical resistance below a critical temperature. These systems utilize Josephson junctions, which are weak links between two superconductors, to create qubits whose states are controlled by manipulating the phase difference across the junction. This precise control allows for intricate quantum manipulations. The coherence times—the duration a qubit maintains its quantum state—in superconducting qubits have significantly improved in recent years, approaching timescales relevant for practical computation. However, these systems often require cryogenic temperatures close to absolute zero.

Trapped ions, while not strictly solid-state systems, find application in solid-state-based quantum computers. These ions, trapped using electromagnetic fields within a solid-state structure, are controlled using lasers. Their internal energy levels serve as the qubits. The scalability of this approach poses a challenge; integrating many trapped ions is complex. Nevertheless, the exquisite control afforded by lasers makes trapped ion qubits highly accurate and offers long coherence times. They provide a powerful platform for quantum simulation as well, allowing researchers to model complex quantum systems and study their behavior.

Another approach uses the electron spin of individual atoms or defects within a solid as qubits. Nitrogen-vacancy (NV) centers in diamond, for example, are point defects where a nitrogen atom replaces a carbon atom in the diamond lattice, creating a vacancy next to it. The electron spin of the NV center, with its unique energy levels, can be controlled and measured using optical and microwave techniques. This system offers the advantage of operation at room temperature, a considerable practical benefit. However, the coherence times are typically shorter than those of superconducting qubits, necessitating further research for improvement.

Quantum dot qubits provide yet another promising pathway. Quantum dots are nanoscale semiconductor structures that confine electrons, creating discrete energy levels that can be used as qubits. The electron spin or the charge state of the quantum dot can serve as the qubit. The advantage of quantum dot qubits lies in their potential for scalability and integration within existing semiconductor technology. However, controlling the interactions between multiple quantum dots, crucial for performing quantum logic operations, requires considerable ingenuity.

Despite the challenges, the pursuit of solid-state quantum computing remains vigorous. Researchers are actively exploring diverse materials and techniques to improve qubit coherence, scalability, and controllability. Each approach faces its own set of obstacles, but advancements continue. The development of efficient error correction codes is crucial, as quantum computations are inherently susceptible to errors from environmental interactions. The eventual success of solid-state quantum computing holds the potential to revolutionize fields from medicine and materials science to finance and artificial intelligence, allowing us to tackle problems currently beyond our reach. The journey is long, but the potential rewards are immense. The integration of these different approaches could potentially lead to hybrid quantum computing architectures, leveraging the strengths of each solid-state system for improved overall performance. This interdisciplinary approach requires extensive collaboration and is likely to drive future breakthroughs. One of the most significant hurdles remains the development of efficient methods to fabricate and scale these quantum systems. Further improvements in material science and nanofabrication techniques are essential to build large-scale quantum computers, capable of demonstrating quantum advantage. The continued exploration and refinement of solid-state-based qubits promises to unlock the unprecedented power of quantum computation and usher in a new era of technological advancement. The future holds remarkable potential for solid-state systems to shape the landscape of quantum computing.

7.2 Strongly Correlated Systems

Strongly correlated electron systems represent a fascinating frontier in condensed matter physics, where the interplay between electrons becomes so significant that it dictates material properties in a way that transcends simple models. Unlike weakly interacting systems, where individual electrons can be treated relatively independently, the strong Coulomb repulsion between electrons in these materials profoundly alters their behavior, leading to emergent phenomena not readily predictable from individual electron characteristics. This complex interplay is at the heart of many intriguing material properties, making them prime candidates for both fundamental research and technological applications.

Understanding these systems requires moving beyond the independent electron approximation, a cornerstone of simpler models. In materials exhibiting strong correlations, the electrons are deeply intertwined, their movements and energies inextricably linked. This intricate dance of interacting electrons creates collective states, leading to emergent phenomena such as high-temperature superconductivity, colossal magnetoresistance, and heavy fermion behavior. These are not merely subtle deviations from simple models; they are entirely new, qualitatively different forms of behavior arising from the collective nature of the interacting electrons.

One crucial aspect of strongly correlated materials is the concept of electron localization. In some cases, strong Coulomb repulsion can lead to the electrons becoming localized in space, preventing them from freely moving through the material. This localization dramatically affects the electrical transport properties, often resulting in insulating behavior even when the density of electrons would suggest metallic conductivity. This phenomenon is particularly intriguing because it highlights the limitations of simple band theory, which relies on the assumption of independent electrons moving in a periodic potential. In strongly correlated materials, electron-electron interactions fundamentally alter this picture.

The Hubbard model, a simplified yet powerful theoretical model, is frequently used to capture the essence of strong correlations. It incorporates the kinetic energy of electrons hopping between lattice sites and the on-site Coulomb repulsion between electrons occupying the same site. The seemingly simple form of this model belies its remarkable complexity, giving rise to rich phase diagrams with diverse behaviors, including metal-insulator transitions and antiferromagnetism. Solving the Hubbard model exactly is a formidable challenge, even for small systems. Approximation techniques such as mean-field theory, dynamical mean-field theory (DMFT), and quantum Monte Carlo methods have been developed to tackle this complex problem.

Experimental probes of strongly correlated systems are equally crucial. Techniques such as angle-resolved photoemission spectroscopy (ARPES), nuclear magnetic resonance (NMR), and neutron scattering provide invaluable insights into the electronic structure, magnetic properties, and collective excitations of these materials. ARPES, for example, can directly map the energy-momentum dispersion of the electrons, revealing the impact of strong correlations on the electronic band structure. NMR and neutron scattering provide complementary information on the local magnetic moments and their dynamics. Combining theoretical models with these experimental techniques allows for a more comprehensive understanding of these complex materials.

The study of strongly correlated systems is not merely an academic pursuit; it has significant technological implications. High-temperature superconductors, for instance, offer the potential for energy-efficient power transmission and high-speed magnetic levitation. Materials exhibiting colossal magnetoresistance have found applications in magnetic field sensors and data storage. Heavy fermion compounds possess unique thermodynamic properties that could be exploited in advanced cooling technologies. Further exploration of these complex materials holds the key to unlocking their full potential, paving the way for groundbreaking advancements in various technological fields. Furthermore, the study of strongly correlated systems pushes the boundaries of our theoretical understanding of quantum many-body problems. The challenges posed by these materials have spurred the development of innovative theoretical techniques and computational methods. The insights gained from these efforts have broader implications for various areas of physics, including quantum field theory and quantum information science.

The ongoing research into strongly correlated systems continues to reveal unexpected and profound phenomena. Unraveling the mysteries of these materials promises not only to advance our understanding of fundamental physics but also to drive technological innovation. The complex interplay between electrons in these materials continues to challenge our intuition and inspire the development of new theoretical and experimental tools. As researchers delve deeper into these intriguing systems, we can expect further breakthroughs, both in our fundamental understanding and in potential technological applications. Their unusual properties often arise from subtle interactions between quantum degrees of freedom, making them a rich source of future research and development. Understanding these systems represents a crucial step toward harnessing the remarkable capabilities of quantum matter. The field is dynamic and ripe with opportunities for significant advancements, making it a pivotal area of study for aspiring physicists. The complexities demand innovative theoretical and experimental approaches, making it a stimulating area for collaborative and interdisciplinary research.

7.3 Future Directions

The quest to understand condensed matter systems remains a vibrant and dynamic field, constantly pushing the boundaries of our knowledge. Unraveling the intricate interplay of classical and quantum phenomena continues to yield surprising discoveries and opens doors to technological advancements we can only begin to imagine. For instance, the ongoing investigation into topological insulators, materials exhibiting unique surface states with robust spin-momentum locking, is paving the way for spintronics applications and potentially revolutionizing data storage and processing. Their inherent topological protection renders them impervious to certain types of scattering, thereby enhancing conductivity and offering unparalleled stability. Further research promises to unearth new topological materials with even more remarkable properties, pushing the limits of what's considered possible in electronic devices.

Our understanding of strongly correlated electron systems, where electron-electron interactions dominate the behavior of materials, is still in its infancy. These systems often defy conventional theoretical approaches, exhibiting phenomena like high-temperature superconductivity and colossal magnetoresistance, which remain largely unexplained. Advanced computational techniques, combined with innovative experimental methodologies, are slowly peeling back the layers of complexity. However, a comprehensive theoretical framework that accounts for the collective behavior of strongly interacting electrons remains an elusive goal. Future progress may involve developing more sophisticated theoretical models incorporating novel concepts and utilizing high-performance computing to tackle the immense computational challenges involved. This could unlock a deeper understanding of these materials, potentially leading to the discovery of new functionalities and technological applications.

The development of quantum computers is another exciting area poised for significant growth. Leveraging the unique properties of condensed matter systems to build qubits, the fundamental building blocks of quantum computers, promises revolutionary advancements in computation. Researchers are exploring various solid-state systems, including superconducting circuits, trapped ions, and quantum dots, as potential platforms for realizing scalable and fault-tolerant quantum computers. The success of these endeavors hinges on improving coherence times and developing efficient qubit control and manipulation techniques. Overcoming these technological hurdles could unlock previously inaccessible computational power, impacting fields ranging from materials science to drug discovery and cryptography.

Beyond these specific areas, the future of condensed matter physics will undoubtedly involve interdisciplinary collaborations, drawing upon expertise from various fields such as mathematics, chemistry, and computer science. This convergence of disciplines will fuel the development of novel theoretical frameworks, cutting-edge experimental techniques, and powerful computational tools, propelling further advancements in the field. The synergy between theory and experiment will continue to be critical, with theoretical predictions guiding experimental design and experimental discoveries informing the refinement of theoretical models. This cyclical feedback loop will drive continuous progress and innovation.

Furthermore, exploring the exotic properties of materials under extreme conditions, such as ultra-high pressures or extremely low temperatures, promises to unveil fundamentally new phenomena. These investigations may reveal previously hidden phases of matter with unexpected properties, extending our knowledge of the quantum world and potentially leading to the discovery of new materials with technological significance. The development of advanced experimental techniques capable of probing materials under such extreme conditions will be paramount to this endeavor. We might witness a paradigm shift in our understanding of fundamental physics and its implications for technology, driven by these investigations.

Finally, the quest for room-temperature superconductivity remains a Holy Grail for physicists and materials scientists alike. While high-Tc superconductors have been discovered, they still require cryogenic cooling, limiting their practical applications. The discovery of a material exhibiting superconductivity at room temperature would revolutionize energy transmission, transportation, and countless other technologies. Achieving this goal requires a deep understanding of the underlying mechanisms governing superconductivity and a strategic approach to materials design and synthesis. Computational techniques will play an increasingly vital role in guiding the experimental search, helping to identify promising candidates for room-temperature superconductivity. This remains one of the most significant challenges and potential breakthroughs in the field of condensed matter physics. The potential benefits of this discovery are immense, potentially reshaping our technological landscape in unforeseen ways.

www.ingramcontent.com/pod-product-compliance
Lightning Source LLC
LaVergne TN
LVHW022020200926
848481LV00007B/2461